An Annotated Guide to Models and Algorithms for Energy Calculations Relating to HVAC Equipment

Prepared by

G.K. YUILL AND ASSOCIATES LTD.

An Annotated Guide to Models and Algorithms for Energy Calculations Relating to HVAC Equipment

 American Society of Heating, Refrigerating, and Air-Conditioning Engineers, Inc.

ISBN 0-910110-73-5

Contents

1. INTRODUCTION

1.1 OBJECTIVE

A large number of models and algorithms that simulate the performance of HVAC equipment have been developed for use in computer programs. Designers of such programs often expend significant amounts of time in searching for this technical information. Therefore, the objective of this project is the preparation of an annotated guide to literature that contains mathematical models and computer algorithms for calculating the performance of HVAC equipment. The models and algorithms that have been described in the literature are categorized according to equipment type and characterized as to approach and limits of applicability.

This annotated guide, which is a stand-alone document that facilitates access to these published algorithms and models, is the first step in ASHRAE's development of a series of publications providing technical information on how to simulate HVAC system components. Although these publications will be aimed at assisting engineers who are attempting to develop their own HVAC system simulation software, they will also be useful to others for different purposes.

This guide and the component simulation guides that follow it will replace "Procedures for Simulating the Performance of Components and Systems for Energy Calculations" (Energy Calculations 2), which was published by ASHRAE in 1976 (ASHRAE 1976).

1.2 SCOPE OF THE ANNOTATED GUIDE

Although the annotated guide is primarily aimed at practicing engineers who wish to develop their own simulation programs for analyzing the energy usage of buildings, it is also aimed at those who wish to carry out dynamic analyses of equipment performance and those who wish to carry out detailed, component-by-component simulation for the purpose of optimizing the design of this equipment. Therefore, the focus of the guide is on steady-state and "quasi-steady-state" (see Section 1.3) algorithms and models, but dynamic ones are also considered. Equipment for all types of buildings from residences to large office buildings is included in this annotated guide.

1.3 DEFINITIONS

Several words that are generally given a range of interpretations are defined here so that they can be used with more precision in this guide. These are as follows:

Model A model focuses on the physical laws and relationships governing the process or device to be simulated, not on the computational steps in the simulation. It is usually expressed in mathematical notation and can usually be inverted to treat any of the parameters and variables it contains as the independent ones. A model is developed by analysis from known physical laws or empirical relationships, or is deduced by experiment.

Algorithm An algorithm is a computational technique for the solution of a problem stated in a model. It has a single narrow purpose and consists of a number of specifically defined steps. While a model may specify the relationship between a number of parameters and variables

1

without defining which are the controlling variables in a particular case, an algorithm proceeds from a particular set of specified "input" parameters and variables to a particular set of "output" variables. Thus, a particular model might be embodied in several different algorithms for use when different sets of parameters and variables are known. An algorithm will have a specific range of parameter and variable values for which it is applicable. It can be expressed in English; in mathematical notation; in a graphical representation, such as a flow chart; or in pseudocode, which follows the general format of common programming languages but allows more flexibility in the use of commonly recognized symbols and notations than a particular language.

Steady-state model A steady-state model is applied once per time period (normally an hour) and the input variables used all have values that apply at that particular time. Although the input variables may change from one time period to the next, the results obtained in each time period are independent of the previous values of those variables.

Quasi-steady-state model A quasi-steady-state model is one in which an essentially steady-state calculation is done in each time period, but a value of a variable from the previous time period is used as a starting point. Such models are not used to try to simulate transient phenomena; rather, they are used to allow a sequential calculation to be done in each time period instead of an iterative or simultaneous calculation.

Transient model A transient or dynamic model is one that describes the time dependence of the phenomenon simulated. In an algorithm based on such a model, the controlling variables will include values from previous time steps.

1.4 ATTRIBUTES OF AN ENERGY CALCULATION MODEL

The differences between an energy calculation model and the other types of thermal simulation models are due to the differing objectives of these models. Energy calculation models are applied in an energy calculation computer program, which is used to predict annual energy consumption of the HVAC system in a particular building. The user wishes the program to focus on the elements of the system he can change. Thus, an energy analysis program might contain a model for a chiller that considers it as a unit, with a small number of inputs that will vary from hour to hour and a small number of outputs that will respond to these inputs. In this model, the chiller is described by a small number of coefficients, which might be physically meaningful.

Such a model is in sharp contrast to the one that would be required by a chiller designer. This model might simulate the performance characteristics of the evaporator, condenser, compressor, and other components, along with the properties of the refrigerant as it undergoes a thermodynamic cycle. If chiller controls were being studied, the transient response of each component might have to be simulated. On the other hand, it would rarely be necessary to use such a model to simulate an entire year of chiller performance.

This comparison highlights the distinctive attributes of an energy calculation program. It must be simple enough to make it feasible to simulate an entire year of system performance and possible to run it several times, as part of the building design process. Its input data requirements must be simple enough that the user can afford to collect the required data for every component of the HVAC system to be simulated. It should simulate as units the components that the building designer selects, not the sub-components that are selected by the component designer. It should be accurate enough to make meaningful distinctions between the options that might face the designer of a building or its HVAC system.

A significant attribute of an energy calculation model is its treatment of time dependent phenomena. Transient heat transfer in a building has a significant impact on the energy loads that must be met by the HVAC system, so the models used for building loads simulation must consider thermal transients. On the other hand, transient phenomena have much less impact on the energy performance of the HVAC system, so it can be simulated independently of its previous condition. An hourly time step provides adequate accuracy.

An energy calculation model may require a sequential calculation in cases where the variables can be evaluated in order with each one dependent only on those that have been evaluated previously. Alternately, it may require a simultaneous or iterative calculation in cases where the variables are all interdependent.

This description of the attributes of an energy calculation model is based on today's conditions, particularly on today's computing costs. Fifteen years ago, a model that included such details as the impact of chilled water return temperature on chiller performance might have been considered to be too costly to run for routine energy analysis and therefore was suitable only for research purposes. In the future, as computing costs continue to drop, it may be possible to justify greater precision, and models that are now considered to be appropriate only for research applications may become appropriate for energy calculations as well.

1.5 TYPES OF ALGORITHMS AND MODELS

Four classifications of models have been developed for the annotated guide to distinguish between the general approaches found in the literature. They are 1) *purely empirical*, 2) *semi-empirical*, 3) *semi-theoretical*, and 4) *fundamental principle*. The same classifications have been applied to algorithms based on these types of models.

Each of the following four sections describes one of these approaches, comments on its range of applicability, and provides an example to illustrate the approach.

1.5.1 Purely Empirical Algorithms and Models

Purely empirical algorithms and models use a simple "black box" concept that ignores physical laws. The equipment is treated as a single unit. No information is required about the interactions between the equipment components or about the physical dimensions, properties, and processes that determine the behavior of the equipment.

Analysis of variance, regression analysis, and intuition are used to select input variables that are important in representing equipment performance and to develop the form of the equations that directly relate the input variables to the desired output variables. Regression analysis is used to fit the equations to discrete performance data obtained from the equipment manufacturer, from laboratory tests, or from more detailed models.

The algorithm or model can be used to predict equipment performance within the range of available test data for the particular equipment operating in a specific environment. However, it cannot be used to predict performance outside this range of test data or for other similar types of equipment operating in different environments. Different equation coefficients, or even different equation forms, must be developed for each individual piece of equipment that is simulated. This feature severely restricts the flexibility of this approach in building energy analysis calculations.

To illustrate this approach, consider the simulation of a water-cooled liquid chiller, operating in a particular building located in a dry climate. An inspection of performance test data for the chiller indicates that its power consumption varies only with outdoor air temperature, according to a quadratic polynomial. The three coefficients of the equation are determined using the least squares method. Although the model can simulate the performance of the chiller under the specified conditions, it does not directly account for intermediate parameters, such as the impact of changes in building thermal resistance or thermal mass on the cooling load, that affect the chiller's performance. Thus, another model would need to be developed if the same chiller were placed in a different building.

1.5.2 Semi-Empirical Algorithms and Models

Semi-empirical algorithms and models use a "black box" concept that is supported by physical laws. The equipment is treated as a single unit. No information is required about the interactions between the equipment components or about the physical dimensions and properties that characterize the equipment. However, some knowledge of the physical processes that govern the behavior of the equipment is required.

Analysis of variance, regression analysis, and intuition, aided by a knowledge of general engineering principles, are used to select input variables that are important in representing equipment performance and to develop the form of the equations that directly relate the input variables to the desired output variables. Regression analysis is used to fit the equations to discrete performance data obtained from the equipment manufacturer, from laboratory tests, or from more detailed models.

The algorithm or model can be used to predict equipment performance within the range of available test data for the particular equipment. Performance predictions can also be made for other similar types of equipment operating in similar environments but it cannot be used to predict performance outside the range of available test data. The simplicity and flexibility of these algorithms and models make them particularly well-suited to simulating complex devices (such as chillers) in building energy analysis calculations that use hourly time steps. However, for simpler devices such as heating coils or air mixing systems, it will usually be possible to take a more theoretical approach.

To illustrate this approach, consider the simulation of a water-cooled liquid chiller. From fundamental principles of the operation of vapor-compression machines, it can be deduced that the capacity and COP of the chiller are functions of the temperature of the water leaving the liquid cooler and of the temperature of the water entering the condenser. Also, since chiller capacity modulation can be achieved through processes such as cylinder unloading, the fraction of full-load power consumption must be a function of chiller capacity. Quadratic equations can be developed to relate the capacity and COP to the water temperatures and to relate the fraction of full-load power consumption to the capacity. The least squares method can be used to fit these equations to performance test data. In addition, variables can be normalized by their nominal values so that different chillers of the same type can be simulated using the same equations and coefficients.

1.5.3 Semi-Theoretical Algorithms and Models

Semi-theoretical algorithms and models rely more on fundamental physical laws but may use empirical expressions for evaluating coefficients, or may use the "black box" approach for simulating particular components of the equipment. Information is required about the interactions between the equipment components and about the physical dimensions, properties, and processes that determine the behavior of the equipment.

Semi-empirical models are developed for two or more components based on discrete performance data obtained from the component or equipment manufacturers, from laboratory tests, or from more detailed models. General engineering principles, such as conservation of energy and conservation of mass, are used to develop relations that represent the interactions between the components. In some cases, semi-empirical correlations are used to represent an interaction between components to fill in gaps in the theory or to simplify the solution.

The algorithm or model can be used to predict equipment performance within the range of available test data for the particular configuration of components that comprise the equipment. The semi-empirical model for a particular component can be used to predict the performance for that component within the range of available test data for that component. Also, these semi-empirical models can be used to predict the performance of other similar types of components operating in similar configurations. However, equipment or component performance cannot be predicted outside the range of available test data for the configuration of components that comprise the equipment. This approach permits more simulation flexibility than the semi-empirical approach. This makes it useful for simulating relatively simple devices (such as heating coils) in building energy analysis computer programs that use hourly time steps. However, the increased level of detail in the semi-theoretical approach makes it less suitable for simulating more complex devices because this additional detail may result in the use of too much computer time or may require more input data than the user can provide.

To illustrate this approach, consider the four basic components of a water-cooled liquid chiller: the compressor (and its drive), the condenser, the refrigerant flow-control device, and the liquid cooler. The heat transfer rates in the condenser and liquid cooler can each be represented by a simple energy balance on the water side and by simple heat exchanger theory (LMTD concept).

Equation fitting can be used to determine the overall heat transfer coefficient of each heat exchanger. To simplify the solution, the interaction between these two components can be represented by a semi-empirical correlation for the COP of the chiller. Then, given the cooling load on the liquid cooler, the water flow rates, and the water inlet temperatures, along with the definition of COP, the power consumption of the chiller can be determined, along with the tower load, refrigerant temperatures, and water outlet temperatures. In this case, no information is required about refrigerant properties or about compressor or expansion device characteristics.

1.5.4 Fundamental Principle Algorithms and Models

Fundamental principle or mechanistic algorithms and models do not use a "black box" concept. Instead, they use general principles of thermodynamics, heat transfer, mass transfer, and momentum transfer to predict pressures, temperatures, energy flow rates, and fluid flow rates to, from, and within each component of the equipment. Little or no performance test data are used, since most of the information required is obtained from physical descriptions of the equipment and its components. Sometimes, semi-empirical correlations are used to fill in gaps in the theory, such as for determining the refrigerant to tube-wall heat transfer coefficient in the liquid cooler.

Since few performance data are used as input to the algorithm or model, predictions of equipment and component performance are limited only by the range of applicability of the theories or correlations used in the simulation. In many cases, the algorithm or model is sufficiently detailed to permit studies of equipment dynamics or of retrofits of components. However, for complex devices (such as chillers or boilers), this amount of detail requires numerous inputs, and programs that implement these algorithms or models tend to consume considerable amounts of computer time. Thus, this approach is of limited use for simulating these complex devices in building energy analysis calculations. On the other hand, there are simple devices that are easily described by fundamental equations and are best simulated by a purely theoretical approach. For example, an air mixing system can be described by the laws of conservation of mass and energy, and these equations can easily be implemented in a building energy analysis program that uses hourly time steps.

To illustrate this approach, consider the simulation of a water-cooled liquid chiller. In particular, consider one component: the liquid cooler. Separate energy balances on the refrigerant and water flows through the liquid cooler relate refrigerant and water inlet and outlet thermodynamic states to the heat transfer rate in the liquid cooler. Simple heat exchanger theory (LMTD concept) relates the heat transfer rate in the liquid cooler to the refrigerant evaporation temperature and to the water inlet and outlet temperatures. The overall heat transfer coefficient of the liquid cooler is determined from the dimensions of the liquid cooler tubes, from the thermal conductivity of the tube walls, and from semi-empirical correlations found in the literature for convective heat transfer coefficients inside and outside the tube walls. With this model, the outlet conditions of the two fluid streams can be determined if the following input variables are known: the cooling load on the liquid cooler, the water inlet temperature and flow rate, and the entering refrigerant state and flow rate. In this approach, the refrigerant state and thermal properties leaving the liquid cooler must be predicted because the refrigerant acts as a link between the liquid cooler and the compressor.

1.6 ORGANIZATION OF THE GUIDE

In this annotated guide, annotations are brief because the organization of the guide itself provides the annotations. In this form, the user of the guide can read the front matter of the guide to understand the differences in simulation approaches and their limits of applicability. Then, for a particular equipment type, the user can consult the relevant section of the guide to select the references that pertain to the particular simulation approach of interest. The format used in the guide to present the literature is as follows for each of Sections 2 through 7.7:

a. Equipment Name
b. Equipment Description
c. Relevant Handbook Chapter

    d.   Key References
    e.   Related References

    To aid users of the guide in locating the particular sections they are interested in, equipment names have been ordered to match as closely as possible the established and familiar order used in the Equipment Handbook (ASHRAE 1988). For some equipment types, no relevant references were found, so these types of equipment are excluded from the guide.

    Once a section of interest is found, users can read a description of the equipment type named in that section to clarify which specific pieces of equipment the references cited pertain to.

    So that users of the guide can easily locate more information describing the equipment named, the relevant chapters in the series of ASHRAE handbooks are listed (ASHRAE 1985, 1986, 1987, 1988).

    Users can then consult lists of key references and/or related references about the particular equipment type of interest. In general, these references represent the most current information available on equipment simulation and are those that are well-known and easily accessible to anyone planning to develop a simulation program. However, references containing equipment simulation techniques that are improvements on the present techniques, and/or that were found in locations that can be considered obscure from the perspective of a North American building energy system simulator, have also been included in the guide. Only literature written in English has been reviewed.

    In this guide, a list of key references is subdivided into one or more of the following eight parts: empirical algorithms, empirical models, semi-empirical algorithms, semi-empirical models, semi-theoretical algorithms, semi-theoretical models, fundamental principle algorithms, and fundamental principle models. These subdivisions are based on the definitions of algorithms and models in Section 1.3 of this guide and on the description of types of algorithms and models in Section 1.5 of this guide. In some cases, fewer than eight subdivisions are used because references for all eight types of algorithms and models were not available in the literature.

    Each list of related references consists of literature that does not contain specific models or algorithms. These references provide background information of interest to those who prepare equipment simulation programs and/or contain information on topics such as heat transfer correlations or methods of testing equipment. The references on testing are included because they provide good indications of the type of data usually available for input to energy calculations.

    Although this annotated guide focuses on the development of simulation software for analyzing the energy usage of buildings, it is also aimed at those who wish to carry out transient or dynamic analyses of equipment performance. **To distinguish between references containing steady-state or quasi-steady-state algorithms or models and those containing transient ones, an asterisk (*) has been placed at the end of each reference to transient models or algorithms**.

    A particular set of references that provide the best models and algorithms has not been recommended in the guide. All of the references that have been annotated contain useful models and/or algorithms. The decision of which model or algorithm is best for a particular application depends on the simulation approach the user is considering and on the application. Thus, this decision has been left to the user.

    Keywords for use in supplementary literature searches are not listed explicitly because the guide contains sufficient information in its present form for the user to select keywords independently.

    In the section titled "Fundamentals" (Section 7.8 of the guide), no annotations of the literature are provided. Also, descriptions are not provided to clarify subsection headings because they are self-explanatory. The references in this section all pertain to fundamental-principle techniques used in equipment simulation.

2. AIR-HANDLING EQUIPMENT

2.1 AIR-DIFFUSING EQUIPMENT

2.1.1 Terminal Boxes

Equipment Description:

A terminal box is a factory-made assembly for controlling air distribution. The box does not alter the composition of the treated air passing through it from the distribution system. It manually or automatically fulfills one or more of the following functions: controls the velocity, pressure, or temperature of the air; controls the airflow rate; mixes airstreams of different temperatures or humidities; or mixes air at high velocity and/or high pressure with air from the conditioned space. Components of the box can include one or more of the following parts: a casing, a mixing section, a manual damper, a heat exchanger, an induction section, and a flow rate controller. Terminal boxes can be single- or dual-duct types, reheat types, variable-air-volume types, or ceiling induction types.

Relevant Handbook Chapter:

ASHRAE. 1988. Chapter 2, "Air-diffusing equipment." *ASHRAE handbook—1988 equipment.* Atlanta: American Society of Heating, Refrigerating, and Air-Conditioning Engineers, Inc.

Key References:

Semi-Theoretical Models
Hanby, V.I., and Clarke, J.A. 1987. *Catalogue of available HVAC plant component models in the U.K.* Report to SERC GR/D/07459. University of Strathclyde, Glasgow. Code A12.
Fundamental Principle Algorithms
ASHRAE. 1983. *Simplified energy analysis using the modified bin method.* Atlanta: American Society of Heating, Refrigerating, and Air-Conditioning Engineers, Inc. pp. 4-19–4-24.
Barlow, R.S. 1983. *User's manual and report on DESSIM version 2.0: A Fortran program for simulation of desiccant cooling systems and components.* Mechanical Engineering Department, Stanford University. p. C-14.
DOE-2. 1982. *Engineer's manual, Version 2.1A.* Lawrence Berkeley Laboratory, Berkeley, California, and Los Alamos National Laboratory, Los Alamos, New Mexico. NTIS, Reference DE83004575. pp. IV.53–IV.137.
Kleiser, J.D. 1987. *Modeling internal combustion engine-powered desiccant cooling systems.* M.Sc. Thesis, University of Nevada, Reno.
TRACE. 1986. "System simulation phase documentation." *Trane Air Conditioning Economics Program: User's Manual, Version 500.* The Trane Company, La Crosse, Wisconsin. pp. 1–3.
TrakLoad. 1986. *Energy audit system: Reference manual, Version 3.1.* Morgan Systems Cor poration, Berkeley, California. pp. 4-13–4-25.
Fundamental Principle Models
Clarke, J.A. 1985. *Energy simulation in building design.* Boston: Adam Hilger Ltd. pp. 238–247.*

7

McQuiston, F.C., and Parker, J.D. 1982. *Heating, ventilating, and air conditioning*, second edition. New York: John Wiley and Sons. pp. 28–30.

Related References:

Chalifoux, A.T. 1986. *A generalized means of simulating HVAC distribution systems on a computer*. M.Sc. Thesis, University of Illinois at Urbana–Champaign.

Hanby, V.I., and Clarke, J.A. 1987. *Catalogue of available HVAC plant component models in the U.K.* Report to SERC GR/D/07459. University of Strathclyde, Glasgow. Code A16.

2.2 FANS

Equipment Description:

A fan is a device that creates a pressure difference and causes airflow by using a power-driven rotating impeller. Fans are generally classified as centrifugal types or as axial flow types, depending on the direction of airflow through the impeller. The volume flowrate of air leaving the fan can be controlled by dampers or orifice plates, inlet vanes, a variable-speed drive, or variable-pitch blades.

Relevant Handbook Chapter:

ASHRAE. 1988. Chapter 3, "Fans." *ASHRAE handbook–1988 equipment*. Atlanta: American Society of Heating, Refrigerating, and Air-Conditioning Engineers, Inc.

Key References:

Semi-Empirical Algorithms

ASHRAE. 1976. *Energy calculations 2–Procedures for simulating the performance of components and systems for energy calculations*. Atlanta: American Society of Heating, Refrigerating, and Air-Conditioning Engineers, Inc. pp. 57–58.

ASHRAE. 1983. *Simplified energy analysis using the modified bin method*. Atlanta: American Society of Heating, Refrigerating, and Air-Conditioning Engineers, Inc. pp. 4-15–4-18.

BLAST. 1986. *Building loads analysis and system thermodynamics program: User's manual, Version 3.0*. U.S. Army Construction Eng'g. Research Lab., Champaign, IL. pp. 5-26–5-27.

TRACE. 1986. "Equipment documentation." *Trane Air Conditioning Economics Program: User's Manual, Version 500*. The Trane Company, La Crosse, Wisconsin. pp. 14–15.

TrakLoad. 1986. *Energy audit system: Reference manual, Version 3.1*. Morgan Systems Corporation, Berkeley, California. pp. 3-73–3-75.

Semi-Empirical Models

Clark, D.R. 1985. *HVACSIM+ building systems and equipment simulation program: Reference manual*. NBSIR 84-2996, National Bureau of Standards, Washington, D.C. pp. 31–32.

Hanby, V.I., and Clarke, J.A. 1987. *Catalogue of available HVAC plant component models in the U.K.* Report to SERC GR/D/07459. University of Strathclyde, Glasgow. Code A17.

Hanby, V.I., and Clarke, J.A. 1987. *Catalogue of available HVAC plant component models in the U.K.* Report to SERC GR/D/07459. University of Strathclyde, Glasgow. Code L9.

HAP. 1987. *E20-II–Hourly analysis program: User's manual, Version 1.1, Vol. 4, Energy analysis, Documentation guide*. Carrier Corporation, Syracuse, New York. pp. 4-1–4-3.

Wright, J.A., and Hanby, V.I. 1988. *Fans*. International Energy Agency, Annex 10, System Simulation, Report AN10-880202-05, University of Liege, Belgium.

Related References:

AMCA. 1987. *Fan application manual–Part 1–Fans and systems*. Arlington Heights, Virginia: Air movement and Control Association, Inc.

ASHRAE. 1986. "Laboratory methods of testing fans for rating." *ASHRAE Standard 51-1985*.

Atlanta: American Society of Heating, Refrigerating, and Air-Conditioning Engineers, Inc.

McQuiston, F.C., and Parker, J.D. 1982. *Heating, ventilating, and air conditioning*, second edition. New York: John Wiley and Sons.

TRNSYS. 1981. *A transient system simulation program: Reference manual.* Solar Energy Laboratory, University of Wisconsin–Madison.

2.3 EVAPORATIVE AIR-COOLING EQUIPMENT

2.3.1 Direct Evaporative Air Coolers

Equipment Description:

A direct evaporative air cooler is a device that cools air by the cooling effect of the evaporation of water into an airstream. The water is in direct contact with the airstream to be cooled and evaporates into it.

Relevant Handbook Chapter:

ASHRAE. 1988. Chapter 4, "Evaporative air-cooling equipment." *ASHRAE handbook–1988 equipment*. Atlanta: American Society of Heating, Refrigerating, and Air-Conditioning Engineers, Inc.

Key References:

Semi-Empirical Algorithms
Barlow, R.S. 1983. *User's manual and report on DESSIM version 2.0: A Fortran program for simulation of desiccant cooling systems and components*. Mechanical Engineering Department, Stanford University.
Semi-Theoretical Algorithms
Dastmalchi, B. 1982. *Seasonal simulation of a solar desiccant cooling system*. M.Sc. Thesis, Department of Mechanical Engineering, Colorado State University.
Semi-Theoretical Models
Maclaine-cross, I.L., and Banks, P.J. 1981. "A general theory for wet surface heat exchangers and its application to regenerative evaporative cooling." *Journal of Heat Transfer, ASME Transactions*, Vol. 103, August. pp. 579–585.

2.3.2 Indirect Evaporative Air Coolers

Equipment Description:

An indirect evaporative air cooler is a device that cools air by the cooling effect of the evaporation of water into an airstream. The air to be cooled passes through one side of a heat exchanger while the cooling air and the evaporating water flow over the other side, so no moisture is added to the air being cooled.

Relevant Handbook Chapter:

ASHRAE. 1988. Chapter 4, "Evaporative air-cooling equipment." *ASHRAE handbook–1988 equipment*. Atlanta: American Society of Heating, Refrigerating, and Air-Conditioning Engineers, Inc.

Key References:

Semi-Empirical Algorithms
Howe, R.R. 1983. *Model and performance characteristics of a commercially sized hybrid air*

9

conditioning system which utilizes a rotary desiccant dehumidifier. M.Sc. Thesis, University of Wisconsin–Madison.

Kleiser, J.D. 1987. *Modeling internal combustion engine-powered desiccant cooling systems.* M.Sc. Thesis, University of Nevada, Reno.

Semi-Theoretical Models

Crum, D.R.; Mitchell, J.W.; and Beckman, W.A. 1987. "Indirect evaporative cooler perform ance." *ASHRAE Transactions*, Vol. 93, Part 2.

Kettleborough, C.F., and Hsieh, C.S. 1983. "The thermal performance of the wet surface plastic plate heat exchanger used as an indirect evaporative cooler." *Journal of Heat Transfer, Transactions of the ASME*, Vol. 105, May. pp. 366–373.

2.3.3 Air Washers

Equipment Description:

An air washer is a type of direct evaporative air cooler in which recirculating water is sprayed into the airstream to cool and clean the air. The air is cooled by the cooling effect of the evaporation of water into the airstream. The water is in direct contact with the airstream to be cleaned and cooled and evaporates into it.

Relevant Handbook Chapter:

ASHRAE. 1988. Chapter 4, "Evaporative air-cooling equipment." *ASHRAE handbook–1988 equipment.* Atlanta: American Society of Heating, Refrigerating, and Air-Conditioning Engineers, Inc.

Key References:

Semi-Theoretical Models

McQuiston, F.C., and Parker, J.D. 1982. *Heating, ventilating, and air conditioning*, second edition. New York: John Wiley and Sons.

Threlkeld, J.L. 1970. *Thermal environmental engineering*, second edition. Englewood Cliffs: Prentice-Hall, Inc.

2.4 HUMIDIFIERS

Equipment Description:

A humidifier is a device that adds moisture to an airstream. The moisture can be added by evaporating or atomizing water, or by injecting steam. Evaporation of water is achieved in a process approximating adiabatic saturation for pan type and wetted-element type water humidifiers. In some cases, the pan type is heated by an electric resistance element, or by a steam or hot-water coil, to increase the evaporation rate.

Relevant Handbook Chapter:

ASHRAE. 1988. Chapter 5, "Humidifiers." *ASHRAE handbook–1988 equipment.* Atlanta: American Society of Heating, Refrigerating, and Air-Conditioning Engineers, Inc.

Key References:

Semi-Theoretical Algorithms

Arneodo, P.; Mazza, A.; and Hanby, V.I. 1988. *Humidifiers.* International Energy Agency, Annex 10, System Simulation, Report AN10-880202-04, University of Liege, Belgium.

Clark, D.R. 1985. "Evaporative humidifier." *HVACSIM+ building systems and equipment simulation program: Reference manual.* NBSIR 84-2996, National Bureau of Standards, Washington, D.C. pp. 72–73.

TrakLoad. 1986. *Energy audit system: Reference manual, Version 3.1.* Morgan Systems Corpo ration, Berkeley, California. p. 4–26.

Semi-Theoretical Models

Hanby, V.I., and Clarke, J.A. 1987. *Catalogue of available HVAC plant component models in the U.K.* Report to SERC GR/D/07459. University of Strathclyde, Glasgow. Code A15.

Hanby, V.I., and Clarke, J.A. 1987. *Catalogue of available HVAC plant component models in the U.K.* Report to SERC GR/D/07459. University of Strathclyde, Glasgow. Code L8.

Fundamental Principle Algorithms

Bullock, C.E. 1978. "Residential central heating system humidifier computer simulation and application guidelines." *ASHRAE Transactions*, Vol. 84, Part 1. pp. 799–814.*

Fundamental Principle Models

Clark, D.R. 1985. "Steam spray humidifier." *HVACSIM+ building systems and equipment simulation program: Reference manual.* NBSIR 84-2996, National Bureau of Standards, Washington, D.C. pp. 89–90.

Clarke, J.A. 1985. *Energy simulation in building design.* Boston: Adam Hilger Ltd. pp. 234–274.*

Hanby, V.I., and Clarke, J.A. 1987. *Catalogue of available HVAC plant component models in the U.K.* Report to SERC GR/D/07459. University of Strathclyde, Glasgow. Code A6.*

Related References:

Chalifoux, A.T. 1986. *A generalized means of simulating HVAC distribution systems on a computer.* M.Sc. Thesis, University of Illinois at Urbana–Champaign.

McQuiston, F.C., and Parker, J.D. 1982. *Heating, ventilating, and air conditioning*, second edition. New York: John Wiley and Sons.

2.5 AIR-COOLING AND DEHUMIDIFYING COILS

2.5.1 Water Coils

Equipment Description:

An air-cooling and dehumidifying water coil is a heat exchanger, with or without extended surfaces, that provides sensible cooling of an airstream under forced circulation, with or without accompanying dehumidification. The fluid that is passed through the coil tubes as a cooling source can be chilled water, brine, or ethylene glycol solution.

Relevant Handbook Chapter:

ASHRAE. 1988. Chapter 6, "Air-cooling and dehumidifying coils." *ASHRAE handbook–1988 equipment.* Atlanta: American Society of Heating, Refrigerating, and Air-Conditioning Engineers, Inc.

Key References:

Semi-Theoretical Algorithms

ASHRAE. 1976. *Energy calculations 2–Procedures for simulating the performance of components and systems for energy calculations.* Atlanta: American Society of Heating, Refrigerating, and Air-Conditioning Engineers, Inc. pp. 43–49.

ASHRAE. 1983. *Simplified energy analysis using the modified bin method.* Atlanta: American Society of Heating, Refrigerating, and Air-Conditioning Engineers, Inc. pp. 4-9–4-14.

ASHRAE. 1988. *ASHRAE handbook–1988 equipment.* Atlanta: American Society of Heating, Refrigerating, and Air-Conditioning Engineers, Inc. pp. 6.7–6.15.

DOE-2. 1982. *Engineer's manual, Version 2.1A.* Lawrence Berkeley Laboratory, Berkeley,

California, and Los Alamos National Laboratory, Los Alamos, New Mexico. NTIS, Reference DE83004575. pp. IV.11–IV.14.

HAP. 1987. *E20-II–Hourly analysis program: User's manual, Version 1.1, Vol. 3, Design load analysis, Documentation guide.* Carrier Corporation, Syracuse, New York. p. 2–6.

Semi-Theoretical Models

Anderson, S.W. 1970. "Air-cooling and dehumidifying coil performance based on ARI Industrial Standard 410–64." *ASHRAE Symposium Bulletin, Heat and Mass Transfer to Extended Surfaces.*

ARI. 1987. "Forced-circulation air-cooling and air-heating coils." *ARI Standard 410–87.* Arlington: Air-Conditioning and Refrigeration Institute.

Hanby, V.I., and Clarke, J.A. 1987. *Catalogue of available HVAC plant component models in the U.K.* Report to SERC GR/D/07459. University of Strathclyde, Glasgow. Code A13.

Hirsch, J.J. 1982. "Simulation of HVAC equipment in the DOE-2 program." *Proceedings of the First International Conference on System Simulation in Buildings.* Liege, Belgium. pp. 96–99.

TRNSYS. 1981. *A transient system simulation program: Reference manual.* Solar Energy Laboratory, University of Wisconsin–Madison. pp. 4.32-1–4.32-6.

Fundamental Principle Algorithms

Clarke, J.A. 1985. *Energy simulation in building design.* Boston: Adam Hilger Ltd. pp. 248–249.

Elmahdy, A.H., and Mitalas, G.P. 1977. *Fortran IV program to simulate cooling and dehumidifying finned-tube multi-row heat exchangers, Computer Program No. 43.* Division of Building Research, National Research Council of Canada, Ottawa.

Ganesh, R. 1987. *A microcomputer model for the performance simulation of cooling and heating coils with applications to heating, ventilating and air-conditioning systems.* Ph.D. Dissertation, University of Missouri–Rolla.

Hanby, V.I., and Clarke, J.A. 1987. *Catalogue of available HVAC plant component models in the U.K.* Report to SERC GR/D/07459. University of Strathclyde, Glasgow. Code A1.

Hanby, V.I., and Clarke, J.A. 1987. *Catalogue of available HVAC plant component models in the U.K.* Report to SERC GR/D/07459. University of Strathclyde, Glasgow. Code L7.

Holmes, M.J. 1982. "The simulation of heating and cooling coils for performance analysis." *Proceedings of the First International Conference on System Simulation in Buildings.* Liege, Belgium. pp. 262–267.

Krempetz, K.J. 1978. *Simulating the performance of cooling and dehumidifying coils.* M.Sc. Thesis, University of Illinois at Urbana–Champaign.

Kusuda, T. 1960. "Coil performance solutions without trial and error." *Air Conditioning, Heating and Ventilating,* Vol. 57, No. 1, January. pp. 73–80.

Fundamental Principle Models

Braun, J.E.; Klein, S.A.; and Mitchell, J.W. 1987. *Effectiveness models for cooling towers and cooling coils.* Solar Energy Laboratory, University of Wisconsin–Madison. pp. 10–19.

Chandrashekar, S., and Green, G.H. 1970. "Dynamic study of a chill water cooling and dehumidifying coil." *ASHRAE Transactions,* Vol. 76, Part 2. pp. 36–51.*

Clark, D.R. 1985. *HVACSIM+ building systems and equipment simulation program: Reference manual.* NBSIR 84-2996, National Bureau of Standards, Washington, D.C. pp.63–68.*

Elmahdy, A.H. 1975. *Analytical and experimental multi-row, finned-tube heat exchanger performance during cooling and dehumidifying processes.* Ph.D. Thesis, Carleton University, Ottawa.*

Elmahdy, A.H., and Mitalas, G.P. 1977. "A simple model for cooling and dehumidifying coils for use in calculating the energy requirements for buildings." *ASHRAE Transactions,* Vol. 83, Part 2. pp. 103–117.

Elmahdy, A.H., and Biggs, R.C. 1978. *Performance simulation of multi-row dry (and/or wet) heat exchangers.* DBR Paper No. 797, National Research Council of Canada, Ottawa.*

Hanby, V.I., and Clarke, J.A. 1987. *Catalogue of available HVAC plant component models in the U.K.* Report to SERC GR/D/07459. University of Strathclyde, Glasgow. Code A2.*

Holmes, M.J. 1988. *Heating and cooling coils.* International Energy Agency, Annex 10, System Simulation, Report AN10-880613-01, University of Liege, Belgium.*

Kusuda, T. 1969. "Effectiveness method for predicting the performance of finned tube coils." *ASHRAE Symposium Bulletin, Heat and Mass Transfer to Extended Surfaces.* New York:

American Society of Heating, Refrigerating, and Air-Conditioning Engineers, Inc.

Maver, T.W., and Clarke, J.A. 1987. *Development of a prototypical component-based energy modelling system*. ABACUS Final Report for Grant GR/D/26610. University of Strathclyde, Glasgow.*

McCullagh, K.R.; Green, G.H.; and Chandrasekar, S. 1969. "An analysis of chill water cooling dehumidifying coils using dynamic relations." *ASHRAE Transactions*, Vol. 75, Part 2. pp. 200–209.*

McQuiston, F.C., and Parker, J.D. 1982. *Heating, ventilating, and air conditioning*, second edition. New York: John Wiley and Sons. pp. 507–519.

Myers, R.J. 1967. *The effect of dehumidification on the air side heat transfer coefficient for a finned-tube coil*. M.Sc. Thesis, University of Minnesota, Minneapolis.

Stoecker, W.F. 1958. *Refrigeration and air conditioning*. New York: McGraw-Hill Book Company. pp. 272–285.

Tamm, H., and Green, G.H. 1970. "Dynamic response relations for combined heat and mass transfer in multi-row crossflow heat exchangers." *International Institute of Refrigeration, Commissions II and III. London. Annex 1970-1*.*

Threlkeld, J.L. 1970. *Thermal environmental engineering*, second edition. Englewood Cliffs, New Jersey: Prentice-Hall, Inc. pp. 254–270.

TrakLoad. 1986. *Energy audit system: Reference manual, Version 3.1*. Morgan Systems Corporation, Berkeley, California. p. 4-25.

Zhang, X., and Warren, M.L. 1988. "Use of a general control simulation program." *ASHRAE Transactions*, Vol. 94, Part 1.*

Related References:

ASHRAE. 1978. "Methods of testing forced circulation air cooling and air heating coils." *ASHRAE Standard 33–78*. Atlanta: American Society of Heating, Refrigerating, and Air-Conditioning Engineers, Inc.

ASHRAE. 1984. "Methods of testing for rating room fan-coil air conditioners." *ASHRAE Standard 79–1984*. Atlanta: American Society of Heating, Refrigerating, and Air-Conditioning Engineers, Inc.

Chalifoux, A.T. 1986. *A generalized means of simulating HVAC distribution systems on a computer*. M.Sc. Thesis, University of Illinois at Urbana–Champaign.

Elmahdy, A.H., and Biggs, R.C. 1979. "Finned tube heat exchanger: Correlation of dry surface heat transfer data." *ASHRAE Transactions*, Vol. 85, Part 2. pp. 262–273.

McQuiston, F.C. 1975. "Fin efficiency with combined heat and mass transfer." *ASHRAE Transactions*, Vol. 81, Part 1. pp. 350–355.

McQuiston, F.C. 1978. "Heat, mass and momentum transfer data for five plate-fin-tube heat transfer surfaces." *ASHRAE Transactions*, Vol. 84, Part 1. pp. 266–293.

McQuiston, F.C. 1978. "Correlation of heat, mass and momentum transport coefficients for plate-fin-tube heat transfer surfaces with staggered tubes." *ASHRAE Transactions*, Vol. 84, Part 1. pp. 294–309.

McQuiston, F.C. 1981. "Finned tube heat exchangers: State-of-the-art for the air side." *ASHRAE Transactions*, Vol. 87, Part 1. pp. 1077–1085.

2.5.2 Direct-Expansion Coils

Equipment Description:

An air-cooling and dehumidifying direct-expansion coil is a heat exchanger, with or without extended surfaces, that provides sensible cooling of an airstream under forced circulation, with or without accompanying dehumidification. The fluid that is passed through the coil tubes as a cooling source is a volatile refrigerant.

Relevant Handbook Chapter:

ASHRAE. 1988. Chapter 6, "Air-cooling and dehumidifying coils." *ASHRAE handbook-1988 equipment*. Atlanta: American Society of Heating, Refrigerating, and Air-Conditioning Engineers, Inc.

Key References:

Semi-Theoretical Algorithms
ASHRAE. 1976. *Energy calculations 2–Procedures for simulating the performance of components and systems for energy calculations*. Atlanta: American Society of Heating, Refrigerating, and Air-Conditioning Engineers, Inc. pp. 43–49.
ASHRAE. 1983. *Simplified energy analysis using the modified bin method*. Atlanta: American Society of Heating, Refrigerating, and Air-Conditioning Engineers, Inc. pp. 4-9–4-14.
ASHRAE. 1988. *ASHRAE handbook–1988 equipment*. Atlanta: American Society of Heating, Refrigerating, and Air-Conditioning Engineers, Inc. pp. 6.7–6.15.
DOE-2. 1982. *Engineer's manual, Version 2.1A*. Lawrence Berkeley Laboratory, Berkeley, California and Los Alamos National Laboratory, Los Alamos, New Mexico. NTIS, Reference DE83004575. pp. IV.11–IV.14.
HAP. 1987. *E20-II–Hourly analysis program: User's manual, Version 1.1, Vol. 3, Design load analysis, Documentation guide*. Carrier Corporation, Syracuse, New York. p. 2–6.
Stamper, E., and Koral, R.L. 1979. *Handbook of air conditioning, heating, and ventilating*. Third Edition. New York: Industrial Press, Inc. pp. 4-5–4-8.

Semi-Theoretical Models
Anderson, S.W. 1970. "Air-cooling and dehumidifying coil performance based on ARI Industrial Standard 410–64." *ASHRAE Symposium Bulletin, Heat and Mass Transfer to Extended Surfaces*.
ARI. 1987. "Forced-circulation air-cooling and air-heating coils." *ARI Standard 410–87*. Arlington: Air-Conditioning and Refrigeration Institute.
Hirsch, J.J. 1982. "Simulation of HVAC equipment in the DOE-2 program." *Proceedings of the First International Conference on System Simulation in Buildings*. Liege, Belgium. pp. 96–99.
Jones, J.W.; Jones, C.D.; Sepsy, C.F.; and Crall, G.C. 1975. "Simulation of a dual duct, reheat air-handling system." *ASHRAE Transactions*, Vol. 81, Part 1. pp. 463–474.

Fundamental Principle Algorithms
Ellison, R.D., and Creswick, F.A. 1978. *A computer simulation of steady-state performance of air-to-air heat pumps*. ORNL/CON-16, Oak Ridge National Laboratory.
Ganesh, R. 1987. *A microcomputer model for the performance simulation of cooling and heating coils with applications to heating, ventilating and air-conditioning systems*. Ph.D. Dissertation, University of Missouri–Rolla.
Holmes, M.J. 1982. "The simulation of heating and cooling coils for performance analysis." *Proceedings of the First International Conference on System Simulation in Buildings*. Liege, Belgium. pp. 262–267.
Krakow, K.I., and Lin, S. 1983. "A computer model for the simulation of multiple source heat pump performance." *ASHRAE Transactions*, Vol. 89, Part 2A. pp. 590–616.
Krempetz, K.J. 1978. *Simulating the performance of cooling and dehumidifying coils*. M.Sc. Thesis, University of Illinois at Urbana–Champaign.
Turaga, M.; Lin, S.; and Fazio, P.P. 1984. "Simulation method for the performance of direct expansion air cooling and dehumidifying coils." *Proceedings of the Workshop on HVAC Controls Modeling and Simulation*. Atlanta.

Fundamental Principle Models
Domanski, P., and Didion, D. 1984. "Mathematical model of an air-to-air heat pump equipped with a capillary tube." *International Journal of Refrigeration*, Vol. 7, No. 4.
Domanski, P.A., and Didion, D.A. 1985. "Simulation of a heat pump operating with a nonazeotropic mixture." *ASHRAE Transactions*, Vol. 91, Part 2B. pp. 1368–1381.
Fischer, S.K., and Rice, C.K. 1983. *The Oak Ridge heat pump models: I. A steady-state computer design model for air-to-air heat pumps*. ORNL/CON-80/R1, Oak Ridge National Laboratory. pp. 47–64.
Goldstein, S.D. 1983. "A mathematically complete analysis of plate-fin heat exchangers." *ASHRAE*

Transactions, Vol. 89, Part 2A. pp. 447–470.

Holmes, M.J. 1988. *Heating and cooling coils*. International Energy Agency, Annex 10, System Simulation, Report AN10-880613-01, University of Liege, Belgium.*

Krakow, K.I., and Lin, S. 1987. "A numerical model of heat pumps having various means of refrigerant flow control and capacity control." *ASHRAE Transactions*, Vol. 93, Part 2.

McQuiston, F.C., and Parker, J.D. 1982. *Heating, ventilating, and air conditioning*, second edition. New York: John Wiley and Sons. pp. 507–519.

Stoecker, W.F. 1958. *Refrigeration and air conditioning*. New York: McGraw-Hill Book Company. pp. 272–285.

Threlkeld, J.L. 1970. *Thermal environmental engineering*, second edition. Englewood Cliffs: Prentice-Hall, Inc. pp. 254–270.

TrakLoad. 1986. *Energy audit system: Reference manual, Version 3.1*. Morgan Systems Corporation, Berkeley, California. p. 4-25.

Related References:

ASHRAE. 1978. "Methods of testing forced circulation air cooling and air heating coils." *ASHRAE Standard 33-78*. Atlanta: American Society of Heating, Refrigerating, and Air-Conditioning Engineers, Inc.

Chalifoux, A.T. 1986. *A generalized means of simulating HVAC distribution systems on a computer*. M.Sc. Thesis, University of Illinois at Urbana–Champaign.

McQuiston, F.C. 1975. "Fin efficiency with combined heat and mass transfer." *ASHRAE Transactions*, Vol. 81, Part 1. pp. 350–355.

McQuiston, F.C. 1981. "Finned tube heat exchangers: State-of-the-art for the air side." *ASHRAE Transactions*, Vol. 87, Part 1. pp. 1077–1085.

2.6 SORPTION DEHUMIDIFICATION EQUIPMENT

Equipment Description:

Sorption dehumidification equipment removes water vapor from an airstream by using liquid or solid sorbent materials that adsorb the water on the surface of the sorbent, or that absorb the water by chemically combining with the water. In regenerative equipment of this type, the mechanism for water removal is reversible. Regeneration of the hygroscopic solution in liquid sorption equipment is achieved by a partial bleed-off and reconcentration of heated solution. Solid sorption equipment regenerates the desiccant with heated air that flows through the desiccant.

Relevant Handbook Chapter:

ASHRAE. 1988. Chapter 7, "Sorption dehumidification and pressure drying equipment." *ASHRAE handbook-1988 equipment*. Atlanta: American Society of Heating, Refrigerating, and Air-Conditioning Engineers, Inc.

Key References:

Semi-Empirical Algorithms

Howe, R.R. 1983. *Model and performance characteristics of a commercially sized hybrid air conditioning system which utilizes a rotary desiccant dehumidifier*. M.Sc. Thesis, University of Wisconsin–Madison.

Kleiser, J.D. 1987. *Modeling internal combustion engine-powered desiccant cooling systems*. M.Sc. Thesis, University of Nevada, Reno.

Semi-Empirical Models

Chen, R.F. 1988. *Assessments of desiccant cooling systems*. M.Sc. Thesis, University of Nevada, Reno.

Turner, R.H.; Kleiser, J.D.; Chen, R.F.; Domingo, N.; and Chen, F. 1988. "Assessment of thermally activated heat pumps with desiccant cooling." *ASHRAE Transactions*, Vol. 94, Part 1.

Van den Bulck, E.; Mitchell, J.W.; and Klein, S.A. 1985. "Design theory for rotary heat and mass exchangers–II. Effectiveness–number-of-transfer-units method for rotary heat and mass exchangers." *International Journal of Heat and Mass Transfer*, Vol. 28, No. 8. pp. 1587–1595.*

Van den Bulck, E.; Mitchell, J.W.; and Klein, S.A. 1986. "The use of dehumidifiers in desiccant cooling and dehumidification systems." *ASME Transactions*, Vol. 108, No. 8 (August). pp. 684–692.*

<u>Fundamental Principle Algorithms</u>

Barlow, R.S. 1982. *Analysis of the adsorption process and of desiccant cooling systems–a pseudo-steady-state model for coupled heat and mass transfer.* Solar Energy Research Institute Report SERI/TR-631-1330, Golden, Colorado.

Barlow, R.S. 1983. *User's manual and report on DESSIM Version 2.0: A Fortran program for simulation of desiccant cooling systems and components.* Mechanical Engineering Department, Stanford University.

<u>Fundamental Principle Models</u>

Dastmalchi, B. 1982. *Seasonal simulation of a solar desiccant cooling system.* M.Sc. Thesis, Department of Mechanical Engineering, Colorado State University. pp. 20–26.

Jurinak, J.J. 1982. *Open cycle solid desiccant cooling–component models and system simulations.* Ph.D. Thesis, University of Wisconsin–Madison.*

Mei, V.C., and Lavan, Z. 1980. "Performance of cross-cooled desiccant dehumidifiers." *Proceedings of the ASME Winter Annual Meeting, Chicago.* New York: American Society of Mechanical Engineers.*

Schultz, K.J., and Mitchell, J.W. 1987. "Experimental analysis of a rotary silica gel dehumidifier." *Proceedings of the ASME Winter Annual Meeting, Boston.* New York: American Society of Mechanical Engineers.*

Van den Bulck, E.; Mitchell, J.W.; and Klein, S.A. 1985. "Design theory for rotary heat and mass exchangers–I. Wave analysis of rotary heat and mass exchangers with infinite transfer coefficients." *International Journal of Heat and Mass Transfer*, Vol. 28, No. 8. pp. 1575–1586.*

Related References:

Crum, D.R.; Mitchell, J.W.; and Beckman, W.A. 1987. "Advanced desiccant cycles." *ASME Conference*, Honolulu, March.

2.7 AIR-HEATING COILS

2.7.1 Steam Coils

Equipment Description:

An air-heating steam coil is a heat exchanger, with or without extended surfaces, that provides sensible heating of an airstream under forced circulation. Steam is passed through the coil tubes as a heating source.

Relevant Handbook Chapter:

ASHRAE. 1988. Chapter 9, "Air-heating coils." *ASHRAE handbook–1988 equipment.* Atlanta: American Society of Heating, Refrigerating, and Air-Conditioning Engineers, Inc.

Key References:

<u>Semi-Theoretical Algorithms</u>
ASHRAE. 1976. *Energy calculations 2–Procedures for simulating the performance of components*

and systems for energy calculations. Atlanta: American Society of Heating, Refrigerating, and Air-Conditioning Engineers, Inc. pp. 40–42.

ASHRAE. 1983. *Simplified energy analysis using the modified bin method*. Atlanta: American Society of Heating, Refrigerating, and Air-Conditioning Engineers, Inc. pp. 4-2–4-8.

DOE-2. 1982. *Engineer's manual, Version 2.1A*. Lawrence Berkeley Laboratory, Berkeley, California, and Los Alamos National Laboratory, Los Alamos, New Mexico. NTIS, Reference DE83004575. pp. IV.101–IV.109.

Semi-Theoretical Models

ARI. 1987. "Forced-circulation air-cooling and air-heating coils." *ARI Standard 410–87*. Arlington: Air-Conditioning and Refrigeration Institute.

Fundamental Principle Algorithms

Ganesh, R. 1987. *A microcomputer model for the performance simulation of cooling and heating coils with applications to heating, ventilating and air-conditioning systems*. Ph.D. Dissertation, University of Missouri–Rolla.

McQuiston, F.C., and Parker, J.D. 1982. *Heating, ventilating, and air conditioning*, second edition. New York: John Wiley and Sons. pp. 469–507.

Fundamental Principle Models

Clark, D.R. 1985. *HVACSIM+ building systems and equipment simulation program: Reference manual*. NBSIR 84-2996, National Bureau of Standards, Washington, D.C. pp. 95–99.*

Holmes, M.J. 1982. "The simulation of heating and cooling coils for performance analysis." *Proceedings of the First International Conference on System Simulation in Buildings*. Liege, Belgium.*

Holmes, M.J. 1988. *Heating and cooling coils*. International Energy Agency, Annex 10, System Simulation, Report AN10-880613-01, University of Liege, Belgium.*

Threlkeld, J.L. 1970. *Thermal environmental engineering*, second edition. Englewood Cliffs, New Jersey: Prentice-Hall, Inc. pp. 235–254.

Related References:

ASHRAE. 1978. "Methods of testing forced circulation air cooling and air heating coils." *ASHRAE Standard 33–78*. Atlanta: American Society of Heating, Refrigerating, and Air-Conditioning Engineers, Inc.

Webb, R.L. 1980. "Air-side heat transfer in finned tube heat exchangers." *Heat Transfer Engineering*, Vol. 1, No. 3. pp. 33–49.

2.7.2 Water Coils

Equipment Description:

An air-heating water coil is a heat exchanger, with or without extended surfaces, that provides sensible heating of an airstream under forced circulation. The fluid that is passed through the coil tubes as a heating source can be water, brine, or ethylene glycol solution.

Relevant Handbook Chapter:

ASHRAE. 1988. Chapter 9, "Air-heating coils." *ASHRAE handbook–1988 equipment*. Atlanta: American Society of Heating, Refrigerating, and Air-Conditioning Engineers, Inc.

Key References:

Semi-Theoretical Algorithms

ASHRAE. 1976. *Energy calculations 2–Procedures for simulating the performance of components and systems for energy calculations*. Atlanta: American Society of Heating, Refrigerating, and Air-Conditioning Engineers, Inc. pp. 40–42.

ASHRAE. 1983. *Simplified energy analysis using the modified bin method*. Atlanta: American

Society of Heating, Refrigerating, and Air-Conditioning Engineers, Inc. pp. 4-2-4-8.

DOE-2. 1982. *Engineer's manual, Version 2.1A*. Lawrence Berkeley Laboratory, Berkeley, California, and Los Alamos National Laboratory, Los Alamos, New Mexico. NTIS, Reference DE83004575. pp. IV.101–IV.109.

Semi-Theoretical Models

ARI. 1987. "Forced-circulation air-cooling and air-heating coils." *ARI Standard 410–87*. Arlington: Air-Conditioning and Refrigeration Institute.

Ashley, C.M. 1946. "A method of analyzing finned coil heat transfer performance." *Refrigerating Engineering*, June. pp. 529–564.

Hamilton, D.C.; Leonard, R.G.; and Pearson, J.T. 1977. "A system model for a discharge air temperature control system." *ASHRAE Transactions*, Vol. 83, Part 1. pp. 251–268.*

Hanby, V.I., and Clarke, J.A. 1987. *Catalogue of available HVAC plant component models in the U.K.* Report to SERC GR/D/07459. University of Strathclyde, Glasgow. Code F2.

Hill, C.R. 1985. "Simulation of a multizone air handler." *ASHRAE Transactions*, Vol. 91, Part 1B. pp. 752–765.*

Jones, J.W.; Jones, C.D.; Sepsy, C.F.; and Crall, G.C. 1975. "Simulation of a dual duct, reheat air-handling system." *ASHRAE Transactions*, Vol. 81, Part 1. pp. 463–474.

Marcev, C.L.; Smith, C.R.; and Bruns, D.D. 1984. "Supervisory control and system optimization of chiller-cooling tower combinations via simulation using primary equipment component models: Part I. Primary equipment component models." *Proceedings of the Workshop on HVAC Control Modeling and Simulation*. Atlanta.

Stoecker, W.F.; Rosario, L.A.; Heidenreich, M.E.; and Phelen, T.R. 1978. "Stability of an air-temperature control loop." *ASHRAE Transactions*, Vol. 84, Part 1. pp. 35–53.*

Fundamental Principle Algorithms

Boot, J.L.; Pearson, J.T.; and Leonard, R.G. 1977. "An improved dynamic response model for finned serpentine cross-flow heat exchangers." *ASHRAE Transactions*, Vol. 83, Part 1. pp. 218–239.*

Ganesh, R. 1987. *A microcomputer model for the performance simulation of cooling and heating coils with applications to heating, ventilating and air-conditioning systems*. Ph.D. Dissertation, University of Missouri–Rolla.

Hanby, V.I., and Clarke, J.A. 1987. *Catalogue of available HVAC plant component models in the U.K.* Report to SERC GR/D/07459. University of Strathclyde, Glasgow. Code L7.

Kusuda, T. 1960. "Coil performance solutions without trial and error." *Air Conditioning, Heating and Ventilating*, Vol. 57, No. 1, January. pp. 73–80.

McCutchan, R.D. 1973. *A simple dynamic model of a finned serpentine heat exchanger*. M.Sc. Thesis, School of Mechanical Engineering, Purdue University.*

McQuiston, F.C., and Parker, J.D. 1982. *Heating, ventilating, and air conditioning*, second edition. New York: John Wiley and Sons. pp. 469–507.

Tobias, J.R. 1973. "Simplified transfer function for temperature response of fluids flowing through coils, pipes or ducts." *ASHRAE Transactions*, Vol. 79, Part 2. pp. 19–22.*

Fundamental Principle Models

Bhargava, S.C.; McQuiston, F.C.; and Zirkle, L.D. 1975. "Transfer functions for crossflow multirow heat exchangers." *ASHRAE Transactions*, Vol. 81, Part 2. pp. 294–314.*

Clark, D.R. 1985. *HVACSIM+ building systems and equipment simulation program: Reference manual*. NBSIR 84-2996, National Bureau of Standards, Washington, D.C. pp. 54–62.*

Clark, D.R.; Hurley, C.W.; and Hill, C.R. 1985. "Dynamic models for HVAC system components." *ASHRAE Transactions*, Vol. 91, Part 1B. pp. 737–750.*

Clarke, J.A. 1985. *Energy simulation in building design*. Boston: Adam Hilger Ltd. pp. 252–254.*

Gartner, J.R., and Harrison, H.L. 1963. "Frequency response transfer functions for a tube in cross-flow." *ASHRAE Transactions*, Vol. 69, Part 2. pp. 323–330.*

Gartner, J.R., and Harrison, H.L. 1965. "Dynamic characteristics of water-to-air crossflow heat exchangers." *ASHRAE Transactions*, Vol. 71, Part 1. pp. 212–224.*

Gartner, J.R., and Daane, L.E. 1969. "Dynamic response relations for a serpentine crossflow heat exchanger with water velocity disturbance." *ASHRAE Transactions*, Vol. 75, Part 1. pp. 53–68.*

Gartner, J.R. 1972. "Simplified dynamic response relations for finned-coil heat exchangers." *ASHRAE Transactions*, Vol. 78, Part 2. pp. 163–169.*

Hanby, V.I., and Clarke, J.A. 1987. *Catalogue of available HVAC plant component models in the U.K.* Report to SERC GR/D/07459. University of Strathclyde, Glasgow. Code A14.

Holmes, M.J. 1982. "The simulation of heating and cooling coils for performance analysis." *Proceedings of the First International Conference on System Simulation in Buildings*. Liege, Belgium.*

Holmes, M.J. 1988. *Heating and cooling coils*. International Energy Agency, Annex 10, System Simulation, Report AN10-880613-01, University of Liege, Belgium.*

Kusuda, T. 1969. "Effectiveness method for predicting the performance of finned tube coils." *ASHRAE Symposium Bulletin, Heat and Mass Transfer to Extended Surfaces*. New York: American Society of Heating, Refrigerating, and Air-Conditioning Engineers, Inc.

McNamara, R.T., and Harrison, H.L. 1967. "A lumped parameter approach to crossflow heat exchanger dynamics." *ASHRAE Transactions*, Vol. 73, Part 2. pp. IV 1.1–IV 1.9.*

Pearson, J.T.; Leonard, R.G.; and McCutchan, R.D. 1974. "Gain and time constant for finned serpentine crossflow heat exchangers." *ASHRAE Transactions*, Vol. 80, Part 2. pp. 255–267.*

Tamm, H. 1969. "Dynamic response relations for multi-row crossflow heat exchangers." *ASHRAE Transactions*, Vol. 75, Part 1. pp. 69–80.*

Threlkeld, J.L. 1970. *Thermal environmental engineering*, second edition. Englewood Cliffs: Prentice-Hall, Inc. pp. 235–254.

TrakLoad. 1986. *Energy audit system: Reference manual, Version 3.1*. Morgan Systems Corporation, Berkeley, California.

Related References:

ASHRAE. 1978. "Methods of testing forced circulation air cooling and air heating coils." *ASHRAE Standard 33–78*. Atlanta: American Society of Heating, Refrigerating, and Air-Conditioning Engineers, Inc.

ASHRAE. 1984. "Methods of testing for rating room fan-coil air conditioners." *ASHRAE Standard 79–1984*. Atlanta: American Society of Heating, Refrigerating, and Air-Conditioning Engineers, Inc.

Chalifoux, A.T. 1986. *A generalized means of simulating HVAC distribution systems on a computer*. M.Sc. Thesis, University of Illinois at Urbana–Champaign.

McQuiston, F.C. 1978. "Heat, mass and momentum transfer data for five plate-fin-tube heat transfer surfaces." *ASHRAE Transactions*, Vol. 84, Part 1. pp. 266–293.

Tamm, H., and Green, G.H. 1973. "Experimental multi-row crossflow heat exchanger dynamics." *ASHRAE Transactions*, Vol. 79, Part 2. pp. 9–18.

Webb, R.L. 1980. "Air-side heat transfer in finned tube heat exchangers." *Heat Transfer Engineering*, Vol. 1, No. 3. pp. 33–49.

3. REFRIGERATION EQUIPMENT

3.1 COMPRESSORS

3.1.1 Positive-Displacement Compressors

Equipment Description:

A positive-displacement compressor is a machine that increases the pressure of refrigerant vapor by reducing the volume of a compression chamber by a fixed amount through work applied to the compression mechanism. The compressor can be a reciprocating type, a rotary (rotating vane, rolling piston, scroll, or trochoidal) type, or a screw type. It can be driven by an engine or an electric motor.

Relevant Handbook Chapter:

ASHRAE. 1988. Chapter 12, "Compressors." *ASHRAE handbook–1988 equipment*. Atlanta: American Society of Heating, Refrigerating, and Air-Conditioning Engineers, Inc.

Key References:

Semi-Theoretical Algorithms
ASHRAE. 1976. *Energy calculations 2–Procedures for simulating the performance of components and systems for energy calculations*. Atlanta: American Society of Heating, Refrigerating, and Air-Conditioning Engineers, Inc. pp. 50–51.
Peitsman, H.C., and Nicolaas, H.J. 1988. *Liquid chilling system*. International Energy Agency, Annex 10, System Simulation, Report AN10-880426-01, University of Liege, Belgium.
Silver, S.C.; Jones, J.W.; Peterson, J.L.; Milbitz, A.; and Hunn, B.D. 1988. *CBS/ICE program user's guide*. Center for Energy Studies, The University of Texas at Austin, Conservation and Solar Research Report No. 7, June. pp. 69–78.
Semi-Theoretical Models
Allen, J.J., and Hamilton, J.F. 1983. "Steady-state reciprocating water chiller models." *ASHRAE Transactions*, Vol. 89, Part 2A. pp. 398–407.
Gluck, R., and Pollak, E. 1978. "Design optimization of air-conditioning systems." *ASHRAE Transactions*, Vol. 84, Part 2. pp. 304–314.
Jones, C.D.; Crall, G.C.; Sepsy, C.F.; and Jones, J.W. 1975. "Simulation and verification of a direct expansion refrigeration system." *ASHRAE Transactions*, Vol. 81, Part 1. pp. 475–483.
Fundamental Principle Algorithms
Ellison, R.D., and Creswick, F.A. 1978. *A computer simulation of steady-state performance of air-to-air heat pumps*. ORNL/CON-16, Oak Ridge National Laboratory.
Krakow, K.I., and Lin, S. 1983. "A computer model for the simulation of multiple source heat pump performance." *ASHRAE Transactions*, Vol. 89, Part 2A. pp. 590–616.
Fundamental Principle Models
Domanski, P., and Didion, D. 1984. "Mathematical model of an air-to-air heat pump equipped with a capillary tube." *International Journal of Refrigeration*, Vol. 7, No. 4.

Domanski, P.A., and Didion, D.A. 1985. "Simulation of a heat pump operating with a nonazeotropic mixture." *ASHRAE Transactions*, Vol. 91, Part 2B. pp. 1368–1381.

Fischer, S.K., and Rice, C.K. 1983. *The Oak Ridge heat pump models: I. A steady-state computer design model for air-to-air heat pumps*. ORNL/CON-80/R1, Oak Ridge National Laboratory.

Fukushima, T.; Takao, K.; and Kawashima, K. 1987. "Mathematical model of a rotary vane type compressor." *ASHRAE Transactions*, Vol. 93, Part 2.*

Krakow, K.I., and Lin, S. 1987. "A numerical model of heat pumps having various means of refrigerant flow control and capacity control." *ASHRAE Transactions*, Vol. 93, Part 2.

McQuiston, F.C., and Parker, J.D. 1982. *Heating, ventilating, and air conditioning*, second edition. New York: John Wiley and Sons. pp. 536–545.

3.1.2 Centrifugal Compressors

Equipment Description:

A centrifugal compressor is a turbomachine that increases the pressure of refrigerant vapor by continuously exchanging angular momentum between a rotating mechanical element and steadily flowing refrigerant. The compressor can be driven by an engine, a gas or steam turbine, or an electric motor.

Relevant Handbook Chapter:

ASHRAE. 1988. Chapter 12, "Compressors." *ASHRAE handbook–1988 equipment*. Atlanta: American Society of Heating, Refrigerating, and Air-Conditioning Engineers, Inc.

Key References:

Semi-Theoretical Algorithms
Peitsman, H.C., and Nicolaas, H.J. 1988. *Liquid chilling system*. International Energy Agency, Annex 10, System Simulation, Report AN10-880426-01, University of Liege, Belgium.
Semi-Theoretical Models
Jackson, W.L.; Chen, F.C.; and Hwang, B.C. 1987. "The simulation and performance of a centrifugal chiller." *ASHRAE Transactions*, Vol. 93, Part 2.
Fundamental Principle Models
Braun, J.E.; Mitchell, J.W.; Klein, S.A.; and Beckman, W.A. 1987. "Models for variable-speed centrifugal chillers." *ASHRAE Transactions*, Vol. 93, Part 1. pp. 1794–1813.
Elder, R.L., and Gill, M.E. 1986. "Simulation of centrifugal compressors for environmental system models." *Proceedings of the Second International Conference on System Simulation in Buildings*. Liege, Belgium.*

3.2 ABSORPTION AIR-CONDITIONING AND REFRIGERATION EQUIPMENT

3.2.1 Lithium Bromide–Water Absorption Equipment

Equipment Description:

A lithium bromide–water absorption machine is a heat-operated refrigeration machine. The absorption cycle uses a secondary fluid to absorb the primary fluid, which is a gaseous refrigerant that has been vaporized in the evaporator. The evaporation process absorbs heat, thus providing the needed refrigerating effect. The refrigerant is driven off from the secondary fluid at a higher pressure, using heat supplied at a higher temperature from an external source. This equipment uses water as the primary fluid (the refrigerant), and lithium bromide as the secondary fluid (the absorber).

Relevant Handbook Chapter:

ASHRAE. 1988. Chapter 13, "Absorption cooling, heating, and refrigeration equipment." *ASHRAE handbook–1988 equipment*. Atlanta: American Society of Heating, Refrigerating, and Air-Conditioning Engineers, Inc.

Key References:

Semi-Empirical Algorithms
BLAST. 1986. *Building loads analysis and system thermodynamics program: User's manual, Version 3.0*. U.S. Army Construction Engineering Research Laboratory, Champaign, Illinois.
Blinn, J.C.; Mitchell, J.W.; and Duffie, J.A. 1979. "Modeling of transient performance of residential solar air-conditioning systems." *Proceedings of the International Solar Energy Society Meeting*. Atlanta.
DOE-2. 1982. *Engineer's manual, Version 2.1A*. Lawrence Berkeley Laboratory, Berkeley, California, and Los Alamos National Laboratory, Los Alamos, New Mexico. NTIS, Reference DE83004575.
Semi-Empirical Models
ASHRAE. 1976. *Energy calculations 2–Procedures for simulating the performance of components and systems for energy calculations*. Atlanta: American Society of Heating, Refrigerating, and Air-Conditioning Engineers, Inc.
TRNSYS. 1981. *A transient system simulation program: Reference manual*. Solar Energy Laboratory, University of Wisconsin–Madison.

Related References:

ASHRAE. 1986. "Methods of testing for rating heat operated unitary air conditioning equipment for cooling." *ASHRAE Standard 40–1986*. Atlanta: American Society of Heating, Refrigerating, and Air-Conditioning Engineers, Inc.
McQuiston, F.C., and Parker, J.D. 1982. *Heating, ventilating, and air conditioning*, second edition. New York: John Wiley and Sons.
Threlkeld, J.L. 1970. *Thermal environmental engineering*, second edition. Englewood Cliffs, New Jersey: Prentice-Hall, Inc.

3.2.2 Ammonia–Water Absorption Equipment

Equipment Description:

An ammonia–water absorption machine is a heat-operated refrigeration machine. The absorption cycle uses a secondary fluid to absorb the primary fluid, which is a gaseous refrigerant that has been vaporized in the evaporator. The evaporation process absorbs heat, thus providing the needed refrigerating effect. The refrigerant is driven off from the secondary fluid at a higher pressure, using heat supplied at a higher temperature from an external source. This equipment uses ammonia as the primary fluid (the refrigerant), and water as the secondary fluid (the absorber).

Relevant Handbook Chapter:

ASHRAE. 1988. Chapter 13, "Absorption cooling, heating, and refrigeration equipment." *ASHRAE handbook–1988 equipment*. Atlanta: American Society of Heating, Refrigerating, and Air-Conditioning Engineers, Inc.

Key References:

Semi-Empirical Algorithms
BLAST. 1986. *Building loads analysis and system thermodynamics program: User's manual, Version 3.0*. U.S. Army Construction Engineering Research Laboratory, Champaign, Illinois.
DOE-2. 1982. *Engineer's manual, Version 2.1A*. Lawrence Berkeley Laboratory, Berkeley, California, and Los Alamos National Laboratory, Los Alamos, New Mexico. NTIS, Reference DE83004575.

Semi-Empirical Models
ASHRAE. 1976. *Energy calculations 2–Procedures for simulating the performance of components and systems for energy calculations*. Atlanta: American Society of Heating, Refrigerating, and Air-Conditioning Engineers, Inc.

Related References:

ASHRAE. 1986. "Methods of testing for rating heat operated unitary air conditioning equipment for cooling." *ASHRAE Standard 40–1986*. Atlanta: American Society of Heating, Refrigerating, and Air-Conditioning Engineers, Inc.

McQuiston, F.C.; and Parker, J.D. 1982. *Heating, ventilating, and air conditioning*, second edition. New York: John Wiley and Sons.

Threlkeld, J.L. 1970. *Thermal environmental engineering*, second edition. Englewood Cliffs, New Jersey: Prentice-Hall, Inc.

3.3 CONDENSERS

3.3.1 Water-Cooled Condensers

Equipment Description:

A water-cooled condenser is a heat exchanger in which refrigerant is condensed by rejecting sensible and latent heat from the refrigerant to a water stream under forced circulation. This heat consists of heat absorbed by the liquid cooler (refrigerant evaporator), plus the heat equivalent of the actual power input required for compression. Water-cooled condensers are usually shell-and-tube types, shell-and-coil types, or tube-in-tube types.

Relevant Handbook Chapter:

ASHRAE. 1988. Chapter 15, "Condensers." *ASHRAE handbook–1988 equipment*. Atlanta: American Society of Heating, Refrigerating, and Air-Conditioning Engineers, Inc.

Key References:

Semi-Theoretical Algorithms
ASHRAE. 1976. *Energy calculations 2–Procedures for simulating the performance of components and systems for energy calculations*. Atlanta: American Society of Heating, Refrigerating, and Air-Conditioning Engineers, Inc. pp. 52–53.

Peitsman, H.C., and Nicolaas, H.J. 1988. *Liquid chilling system*. International Energy Agency, Annex 10, System Simulation, Report AN10-880426-01, University of Liege, Belgium.

Silver, S.C.; Jones, J.W.; Peterson, J.L.; Milbitz, A.; and Hunn, B.D. 1988. *CBS/ICE program user's guide*. Center for Energy Studies, The University of Texas at Austin, Conservation and Solar Research Report No. 7, June. pp. 87–91.

Semi-Theoretical Models
Allen, J.J., and Hamilton, J.F. 1983. "Steady-state reciprocating water chiller models." *ASHRAE Transactions*, Vol. 89, Part 2A. pp. 398–407.

Gluck, R., and Pollak, E. 1978. "Design optimization of air-conditioning systems." *ASHRAE Transactions*, Vol. 84, Part 2. pp. 304–314.

Fundamental Principle Algorithms
Krakow, K.I., and Lin, S. 1983. "A computer model for the simulation of multiple source heat pump performance." *ASHRAE Transactions*, Vol. 89, Part 2A. pp. 590–616.

Fundamental Principle Models
Braun, J.E.; Mitchell, J.W.; Klein, S.A.; and Beckman, W.A. 1987. "Models for variable-speed centrifugal chillers." *ASHRAE Transactions*, Vol. 93, Part 1. pp. 1794–1813.

Jackson, W.L.; Chen, F.C.; and Hwang, B.C. 1987. "The simulation and performance of a centrifugal chiller." *ASHRAE Transactions*, Vol. 93, Part 2.

Krakow, K.I., and Lin, S. 1987. "A numerical model of heat pumps having various means of refrigerant flow control and capacity control." *ASHRAE Transactions*, Vol. 93, Part 2.

Myers, G.E.; Mitchell, J.W.; and Lindeman, Jr., C.F. 1970. "The transient response of heat exchangers having an infinite capacitance rate fluid." *ASME Transactions, Series C, Journal of Heat Transfer*, Vol. 92, May. pp. 269–275.*

Yasuda, H.; Touber, S.; and Machielsen, C.H.M. 1983. "Simulation model of a vapor compression refrigeration system." *ASHRAE Transactions*, Vol. 89, Part 2A. pp. 408–425.*

3.3.2 Air-Cooled Condensers

Equipment Description:

An air-cooled condenser is a heat exchanger in which refrigerant is condensed by rejecting sensible and latent heat from the refrigerant to an airstream under free convection or forced circulation (wind or fan). This heat consists of heat absorbed by the liquid cooler (refrigerant evaporator), plus the heat equivalent of the actual power input required for compression. Air-cooled condensers usually have extended surfaces.

Relevant Handbook Chapter:

ASHRAE. 1988. Chapter 15, "Condensers." *ASHRAE handbook–1988 equipment*. Atlanta: American Society of Heating, Refrigerating, and Air-Conditioning Engineers, Inc.

Key References:

Semi-Theoretical Algorithms

ASHRAE. 1976. *Energy calculations 2–Procedures for simulating the performance of components and systems for energy calculations*. Atlanta: American Society of Heating, Refrigerating, and Air-Conditioning Engineers, Inc. p. 53.

Semi-Theoretical Models

Jones, C.D.; Crall, G.C.; Sepsy, C.F.; and Jones, J.W. 1975. Simulation and verification of a direct expansion refrigeration system." *ASHRAE Transactions*, Vol. 81, Part 1. pp. 475–483.

Fundamental Principle Algorithms

Anand, N.K., and Tree, D.R. 1982. "Steady state simulation of a single tube-finned condenser heat exchanger." *ASHRAE Transactions*, Vol. 88, Part 2. pp. 185–200.

Ellison, R.D., and Creswick, F.A. 1978. *A computer simulation of steady-state performance of air-to-air heat pumps*. ORNL/CON-16, Oak Ridge National Laboratory.

Krakow, K.I., and Lin, S. 1983. "A computer model for the simulation of multiple source heat pump performance." *ASHRAE Transactions*, Vol. 89, Part 2A. pp. 590–616.

Silver, S.C.; Jones, J.W.; Peterson, J.L.; Milbitz, A.; and Hunn, B.D. 1988. *CBS/ICE program user's guide*. Center for Energy Studies, The University of Texas at Austin, Conservation and Solar Research Report No. 7, June. pp. 99–111.

Fundamental Principle Models

Domanski, P., and Didion, D. 1984. "Mathematical model of an air-to-air heat pump equipped with a capillary tube." *International Journal of Refrigeration*, Vol. 7, No. 4.

Domanski, P.A., and Didion, D.A. 1985. "Simulation of a heat pump operating with a non azeotropic mixture." *ASHRAE Transactions*, Vol. 91, Part 2B. pp. 1368–1381.

Ellison, R.D.; Creswick, F.A.; Fischer, S.K.; and Jackson, W.L. 1981. "A computer model for air-cooled refrigerant condensers with specified refrigerant circuiting." *ASHRAE Transactions*, Vol. 87, Part 1. pp. 1106–1124.

Fischer, S.K., and Rice, C.K. 1983. *The Oak Ridge heat pump models: I. A steady-state com puter design model for air-to-air heat pumps*. ORNL/CON-80/R1, Oak Ridge National Laboratory.

Goldstein, S.D. 1983. "A mathematically complete analysis of plate-fin heat exchangers." *ASHRAE*

Transactions, Vol. 89, Part 2A. pp. 447–470.

Krakow, K.I., and Lin, S. 1987. "A numerical model of heat pumps having various means of refrigerant flow control and capacity control." *ASHRAE Transactions*, Vol. 93, Part 2.

Myers, G.E.; Mitchell, J.W.; and Lindeman, Jr., C.F. 1970. "The transient response of heat exchangers having an infinite capacitance rate fluid." *ASME Transactions, Series C, Journal of Heat Transfer*, Vol. 92, May. pp. 269–275.*

3.3.3 Evaporative Condensers

Equipment Description:

An evaporative condenser is a condensing coil that is continuously wetted by a recirculating water stream, while air is simultaneously directed through the coil, evaporating part of the water stream to increase the cooling effect.

Relevant Handbook Chapter:

ASHRAE. 1988. Chapter 15, "Condensers." *ASHRAE handbook–1988 equipment*. Atlanta: American Society of Heating, Refrigerating, and Air-Conditioning Engineers, Inc.

Key References:

Semi-Theoretical Algorithms

Silver, S.C.; Jones, J.W.; Peterson, J.L.; Milbitz, A.; and Hunn, B.D. 1988. *CBS/ICE program user's guide*. Center for Energy Studies, The University of Texas at Austin, Conservation and Solar Research Report No. 7, June.

Webb, R.L., and Villacres, A. 1984. "Algorithms for performance simulation of cooling towers, evaporative condensers, and fluid coolers." *ASHRAE Transactions*, Vol. 90, Part 2. pp. 416–458.

Webb, R.L., and Villacres, A. 1984. "Performance simulation of evaporative heat exchangers (Cooling towers, fluid coolers, and condensers)." *AIChE Symposium Session*, Vol. 79. pp. 458–464.

Semi-Theoretical Models

Korenic, B., and Leidenfrost, W. 1984. "Transient response of evaporative condensers." *Proceedings of the Workshop on HVAC Controls Modeling and Simulation*. Atlanta.*

Maclaine-cross, I.L., and Banks, P.J. 1981. "A general theory for wet surface heat exchangers and its application to regenerative evaporative cooling." *Journal of Heat Transfer, ASME Transactions*, Vol. 103, August. pp. 579–585.

Villacres, A. 1984. *Computer simulation of performance of evaporative heat exchangers*. M.Sc. Thesis, Pennsylvania State University.

3.4 LIQUID COOLERS (EVAPORATORS)

Equipment Description:

A liquid cooler is a heat exchanger in which refrigerant is evaporated by absorbing heat from water or brine under forced circulation. It can be a direct-expansion type, a flooded shell-and-tube type, a shell-and-coil type, or a Baudelot type.

Relevant Handbook Chapter:

ASHRAE. 1988. Chapter 16, "Liquid coolers." *ASHRAE handbook–1988 equipment*. Atlanta: American Society of Heating, Refrigerating, and Air-Conditioning Engineers, Inc.

Key References:

Semi-Theoretical Algorithms

ASHRAE. 1976. *Energy calculations 2–Procedures for simulating the performance of components and systems for energy calculations.* Atlanta: American Society of Heating, Refrigerating, and Air-Conditioning Engineers, Inc. p. 55.

Peitsman, H.C., and Nicolaas, H.J. 1988. *Liquid chilling system.* International Energy Agency, Annex 10, System Simulation, Report AN10-880426-01, University of Liege, Belgium.

<u>Semi-Theoretical Models</u>

Allen, J.J., and Hamilton, J.F. 1983. "Steady-state reciprocating water chiller models." *ASHRAE Transactions*, Vol. 89, Part 2A. pp. 398–407.

Gluck, R., and Pollak, E. 1978. "Design optimization of air-conditioning systems." *ASHRAE Transactions*, Vol. 84, Part 2. pp. 304–314.

<u>Fundamental Principle Algorithms</u>

Krakow, K.I., and Lin, S. 1983. "A computer model for the simulation of multiple source heat pump performance." *ASHRAE Transactions*, Vol. 89, Part 2A. pp. 590–616.

<u>Fundamental Principle Models</u>

Braun, J.E.; Mitchell, J.W.; Klein, S.A.; and Beckman, W.A. 1987. "Models for variable-speed centrifugal chillers." *ASHRAE Transactions*, Vol. 93, Part 1. pp. 1794–1813.

Jackson, W.L.; Chen, F.C.; and Hwang, B.C. 1987. "The simulation and performance of a centrifugal chiller." *ASHRAE Transactions*, Vol. 93, Part 2.

Krakow, K.I., and Lin, S. 1987. "A numerical model of heat pumps having various means of refrigerant flow control and capacity control." *ASHRAE Transactions*, Vol. 93, Part 2.

Myers, G.E.; Mitchell, J.W.; and Lindeman, Jr., C.F. 1970. "The transient response of heat exchangers having an infinite capacitance rate fluid." *ASME Transactions, Series C, Journal of Heat Transfer*, Vol. 92, May. pp. 269–275.*

Yasuda, H.; Touber, S.; and Machielsen, C.H.M. 1983. "Simulation model of a vapor compression refrigeration system." *ASHRAE Transactions*, Vol. 89, Part 2A. pp. 408–425.*

Related References:

Jacobson, D.I. 1986. *Simulation and optimization of the operation of an ice storage system.* M.Sc. Thesis, University of Illinois at Urbana–Champaign.

Silver, S.C.; Jones, J.W.; Peterson, J.L.; Milbitz, A.M.; and Hunn, B.D. 1988. "Component models for simulation of ice storage systems." *ASHRAE RP–459 Technical Paper*, Presented at ASHRAE Winter Meeting at Dallas.

Silver, S.C.; Jones, J.W.; Peterson, J.L.; Milbitz, A.; and Hunn, B.D. 1988. *CBS/ICE program user's guide.* Center for Energy Studies, The University of Texas at Austin, Conservation and Solar Research Report No. 7, June.

3.5 LIQUID CHILLING SYSTEMS

Equipment Description:

A liquid chilling system is a mechanical vapor-compression machine that cools a secondary coolant such as water or brine for air conditioning or refrigeration. Its basic components include a liquid cooler (refrigerant evaporator), a compressor (and its drive), a condenser, and refrigerant flow-control devices.

Relevant Handbook Chapter:

ASHRAE. 1988. Chapter 17, "Liquid chilling systems." *ASHRAE handbook–1988 equipment.* Atlanta: American Society of Heating, Refrigerating, and Air-Conditioning Engineers, Inc.

Key References:

<u>Empirical Models</u>

Bullock, C.E. 1984. "Dynamic simulation models for commercial air conditioning and heat pump systems." *Proceedings of the Workshop on HVAC Controls Modeling and Simulation*. Atlanta.

Semi-Empirical Algorithms

ASHRAE. 1976. *Energy calculations 2–Procedures for simulating the performance of components and systems for energy calculations*. Atlanta: American Society of Heating, Refrigerating, and Air-Conditioning Engineers, Inc. pp. 32–35.

ASHRAE. 1983. *Simplified energy analysis using the modified bin method*. Atlanta: American Society of Heating, Refrigerating, and Air-Conditioning Engineers, Inc. pp. 6-10–6-20.

BLAST. 1986. *Building loads analysis and system thermodynamics program: User's manual, Version 3.0*. U.S. Army Construction Engineering Research Laboratory, Champaign, Illinois. pp. 6-35–6-48.

Clark, C.S. 1987. *A methodology for modeling water chillers in building energy analysis programs*. M.Sc. Thesis, University of Illinois at Urbana–Champaign. pp. 12–16.

DOE-2. 1982. *Engineer's manual, Version 2.1A*. Lawrence Berkeley Laboratory, Berkeley, California, and Los Alamos National Laboratory, Los Alamos, New Mexico. NTIS, Reference DE83004575. pp. V.35–V.45.

Hackner, R.J.; Mitchell, J.W.; and Beckman, W.A. 1984. "HVAC system dynamics and energy use in existing buildings." *ASHRAE Transactions*, Vol. 90, Part 2.

Hanby, V.I., and Clarke, J.A. 1987. *Catalogue of available HVAC plant component models in the U.K.* Report to SERC GR/D/07459. University of Strathclyde, Glasgow. Code A11.

HAP. 1987. *E20-II–Hourly analysis program: User's manual, Version 1.1, Vol. 4, Energy analysis, Documentation guide*. Carrier Corporation, Syracuse, New York. pp. 5-1–5-5.

Leverenz, D.J., and Bergan, N.E. 1983. "Development and validation of a reciprocating chiller model for hourly energy analysis programs." *ASHRAE Transactions*, Vol. 89, Part 1A. pp. 156–174.

TRACE. 1986. "Equipment documentation." *Trane Air Conditioning Economics Program: User's Manual, Version 500*. The Trane Company, La Crosse, Wisconsin. pp. 2–6.

TrakLoad. 1986. *Energy audit system: Reference manual, Version 3.1*. Morgan Systems Corporation, Berkeley, California. pp. 3-77–3-80 and 4-35–4-37.

Semi-Empirical Models

Allen, J.J., and Hamilton, J.F. 1983. "Steady-state reciprocating water chiller models." *ASHRAE Transactions*, Vol. 89, Part 2A. pp. 398–407.

Gajanana, B.C. 1979. *Computer simulation and application studies of a space conditioning system with a diesel driven chiller and dual tank thermal storage*. Ph.D. Thesis, Drexel University, Michigan. pp. 27–32.

Hanby, V.I., and Clarke, J.A. 1987. *Catalogue of available HVAC plant component models in the U.K.* Report to SERC GR/D/07459. University of Strathclyde, Glasgow. Code L5.

TRNSYS. 1981. *A transient system simulation program: reference manual*. Solar Energy Laboratory, University of Wisconsin–Madison. pp. 4.6.9-1–4.6.9-10.

Semi-Theoretical Algorithms

Marcev, C.L.; Smith, C.R.; and Bruns, D.D. 1984. "Supervisory control and system optimization of chiller-cooling tower combinations via simulation using primary equipment component models: Part I. Primary equipment component models." *Proceedings of the Workshop on HVAC Control Modeling and Simulation*. Atlanta.

Semi-Theoretical Models

Jackson, W.L.; Chen, F.C.; and Hwang, B.C. 1987. "The simulation and performance of a centrifugal chiller." *ASHRAE Transactions*, Vol. 93, Part 2.

Fundamental Principle Models

Braun, J.E.; Mitchell, J.W.; Klein, S.A.; and Beckman, W.A. 1987. "Models for variable-speed centrifugal chillers." *ASHRAE Transactions*, Vol. 93, Part 1. pp. 1794–1813.

Related References:

ASHRAE. 1978. "Methods of testing liquid chilling packages." *ASHRAE Standard 30–78*. Atlanta: American Society of Heating, Refrigerating, and Air-Conditioning Engineers, Inc.

3.6 REFRIGERANT-CONTROL EXPANSION DEVICES

Equipment Description:

A refrigerant-control expansion device controls the mass flow rate of refrigerant entering the evaporator so it equals the rate at which refrigerant can be completely vaporized in the evaporator by absorption of heat. The device can be a thermostatic expansion valve, a constant-pressure expansion valve, or a capillary tube.

Relevant Handbook Chapter:

ASHRAE. 1988. Chapter 19, "Refrigerant-control devices." *ASHRAE handbook–1988 equipment.* Atlanta: American Society of Heating, Refrigerating, and Air-Conditioning Engineers, Inc.

Key References:

Semi-Theoretical Models
Krakow, K.I., and Lin, S. 1987. "A numerical model of heat pumps having various means of refrigerant flow control and capacity control." *ASHRAE Transactions*, Vol. 93, Part 2.
Fundamental Principle Algorithms
Silver, S.C.; Jones, J.W.; Peterson, J.L.; Milbitz, A.; and Hunn, B.D. 1988. *CBS/ICE program user's guide.* Center for Energy Studies, The University of Texas at Austin, Conservation and Solar Research Report No. 7, June. pp. 112–115.
Fundamental Principle Models
Domanski, P., and Didion, D. 1984. "Mathematical model of an air-to-air heat pump equipped with a capillary tube." *International Journal of Refrigeration*, Vol. 7, No. 4.
Domanski, P.A., and Didion, D.A. 1985. "Simulation of a heat pump operating with a nonazeotropic mixture." *ASHRAE Transactions*, Vol. 91, Part 2B. pp. 1368–1381.
Fischer, S.K., and Rice, C.K. 1983. *The Oak Ridge heat pump models: I. A steady-state computer design model for air-to-air heat pumps.* ORNL/CON-80/R1, Oak Ridge National Laboratory.
Fischer, S.K.; Rice, C.K.; and Jackson, W.L. 1986. *The Oak Ridge heat pump model: Mark III version program documentation.* ORNL/TM-10192, Oak Ridge National Laboratory.
Yasuda, H.; Touber, S.; and Machielsen, C.H.M. 1983. "Simulation model of a vapor compression refrigeration system." *ASHRAE Transactions*, Vol. 89, Part 2A. pp. 408–425.*

Related References:

Braun, J.E.; Mitchell, J.W.; Klein, S.A.; and Beckman, W.A. 1987. "Models for variable-speed centrifugal chillers." *ASHRAE Transactions*, Vol. 93, Part 1. pp. 1794–1813.

3.7 COOLING TOWERS

3.7.1 Direct-Contact Towers

3.7.1.1 Non-mechanical-draft towers

Equipment Description:

A direct-contact non-mechanical-draft cooling tower is a device whose purpose is to cool water by transferring energy to the atmosphere. It uses the evaporation of water into the air as well as the transfer of sensible heat to the air to achieve its cooling effect. The water to be cooled is in direct contact with the air to which it is transferring energy. Some of this water evaporates into the airstream. A mechanical device (fan) is not used to create an airflow through the tower. Instead, airflow is created by the buoyancy of the warmed moist air leaving the tower, or by the aspirating effect of the water spray.

Relevant Handbook Chapter:

ASHRAE. 1988. Chapter 20, "Cooling towers." *ASHRAE handbook–1988 equipment*. Atlanta: American Society of Heating, Refrigerating, and Air-Conditioning Engineers, Inc.

Key References:

Semi-Empirical Models
Whiller, A. 1976. "A fresh look at the calculation of performance of cooling towers." *ASHRAE Transactions*, Vol. 82, Part 1. pp. 269–282.
Semi-Theoretical Algorithms
Majumdar, A.K.; Singhal, A.K.; and Spalding, D.B. 1983. "VERA2D–A computer program for two-dimensional analysis of flow, heat and mass transfer in evaporative cooling towers: Vol. I–Mathematical formulation, solution procedure, and applications." *EPRI Report, Contract No. RP 1262-1*.
Majumdar, A.K., and Singhal, A.K. 1983. "VERA2D–A computer program for two-dimensional analysis of flow, heat and mass transfer in evaporative cooling towers: Vol. II–user's manual." *EPRI Report No. CS 2923, Contract No. RP 1262-1*.
Semi-Theoretical Models
Braun, J.E.; Klein, S.A.; and Mitchell, J.W. 1987. *Effectiveness models for cooling towers and cooling coils*. Solar Energy Laboratory, University of Wisconsin–Madison.
Cheng, M.S. 1976. *Analysis of different types of dry-wet cooling towers*. Ph.D. Thesis, University of Iowa.
Maclaine-cross, I.L., and Banks, P.J. 1981. "A general theory for wet surface heat exchangers and its application to regenerative evaporative cooling." *Journal of Heat Transfer, ASME Transactions*, Vol. 103, August. pp. 579–585.
Majumdar, A.K.; Singhal, A.K.; and Spalding, D.B. 1983. "Numerical modeling of wet cooling towers–Part 1: Mathematical and physical models." *Journal of Heat Transfer, ASME Transactions*, Vol. 105. pp. 728–735.
Majumdar, A.K.; Singhal, A.K.; Reilly, H.E.; and Bartz, J.A. 1983. "Numerical modeling of wet cooling towers–Part 2: Application to natural and mechanical draft towers." *Journal of Heat Transfer, ASME Transactions*, Vol. 105. pp. 736–743.

3.7.1.2 Mechanical-draft towers

Equipment Description:

A direct-contact mechanical-draft cooling tower is a device whose purpose is to cool water by transferring energy to the atmosphere. It uses the evaporation of water into the air as well as the transfer of sensible heat to the air to achieve its cooling effect. The water to be cooled is in direct contact with the air to which it is transferring energy. Some of this water evaporates into the airstream. A fan is used to circulate air through the tower.

Relevant Handbook Chapter:

ASHRAE. 1988. Chapter 20, "Cooling towers." *ASHRAE handbook–1988 equipment*. Atlanta: American Society of Heating, Refrigerating, and Air-Conditioning Engineers, Inc.

Key References:

Empirical Models
ASHRAE. 1976. *Energy calculations 2–Procedures for simulating the performance of components and systems for energy calculations*. Atlanta: American Society of Heating, Refrigerating, and Air-Conditioning Engineers, Inc.

Gajanana, B.C. 1979. *Computer simulation and application studies of a space conditioning system with a diesel driven chiller and dual tank thermal storage.* Ph.D. Thesis, Drexel University, Michigan.

Semi-Empirical Algorithms

BLAST. 1986. *Building loads analysis and system thermodynamics program: User's manual, Version 3.0.* U.S. Army Construction Engineering Research Laboratory, Champaign, Illinois.

DOE-2. 1982. *Engineer's manual, Version 2.1A.* Lawrence Berkeley Laboratory, Berkeley, California, and Los Alamos National Laboratory, Los Alamos, New Mexico. NTIS, Reference DE83004575.

Peitsman, H.C., and Nicolaas, H.J. 1988. *Liquid chilling system.* International Energy Agency, Annex 10, System Simulation, Report AN10-880426-01, University of Liege, Belgium.

Semi-Empirical Models

Marcev, C.L.; Smith, C.R.; and Bruns, D.D. 1984. "Supervisory control and system optimization of chiller-cooling tower combinations via simulation using primary equipment component models: Part I. Primary equipment component models." *Proceedings of the Workshop on HVAC Control Modeling and Simulation.* Atlanta.

Semi-Theoretical Algorithms

Arneodo, P.; Giaretto, V.; and Mazza, A. 1988. *Cooling tower.* International Energy Agency, Annex 10, System Simulation, Report AN10-880520-02, University of Liege, Belgium.

Majumdar, A.K.; Singhal, A.K.; and Spalding, D.B. 1983. "VERA2D–A computer program for two-dimensional analysis of flow, heat and mass transfer in evaporative cooling towers: Vol. I–Mathematical formulation, solution procedure, and applications." *EPRI Report, Contract No. RP 1262-1.*

Majumdar, A.K., and Singhal, A.K. 1983. "VERA2D–A computer program for two-dimensional analysis of flow, heat and mass transfer in evaporative cooling towers: Vol. II–user's manual." *EPRI Report No. CS 2923, Contract No. RP 1262-1.*

Silver, S.C.; Jones, J.W.; Peterson, J.L.; Milbitz, A.; and Hunn, B.D. 1988. *CBS/ICE program user's guide.* Center for Energy Studies, The University of Texas at Austin, Conservation and Solar Research Report No. 7, June.

Webb, R.L., and Villacres, A. 1984. "Algorithms for performance simulation of cooling towers, evaporative condensers, and fluid coolers." *ASHRAE Transactions,* Vol. 90, Part 2. pp. 416–458.

Webb, R.L., and Villacres, A. 1984. "Performance simulation of evaporative heat exchangers (Cooling towers, fluid coolers, and condensers)." *AIChE Symposium Session,* Vol. 79. pp. 458–464.

Semi-Theoretical Models

Baker, K.L. 1978. *An improved method for cross-flow cooling tower performance analysis.* M.Sc. Thesis, University of Kentucky.

Braun, J.E.; Klein, S.A.; and Mitchell, J.W. 1987. *Effectiveness models for cooling towers and cooling coils.* Solar Energy Laboratory, University of Wisconsin–Madison.

Cheng, M.S. 1976. *Analysis of different types of dry-wet cooling towers.* Ph.D. Thesis, University of Iowa.

Crum, D.R.; Mitchell, J.W.; and Beckman, W.A. 1987. "Indirect evaporative cooler performance." *ASHRAE Transactions,* Vol. 93, Part 2.

Hanby, V.I., and Clarke, J.A. 1987. *Catalogue of available HVAC plant component models in the U.K.* Report to SERC GR/D/07459. University of Strathclyde, Glasgow.

Jalali, E.A. 1979. *Cooling tower analyses and economic evaluation.* M.Sc. Thesis, California State University, Long Beach.

Maclaine-cross, I.L., and Banks, P.J. 1981. "A general theory for wet surface heat exchangers and its application to regenerative evaporative cooling." *Journal of Heat Transfer, ASME Transactions,* Vol. 103, August. pp. 579–585.

Majumdar, A.K.; Singhal, A.K.; and Spalding, D.B. 1983. "Numerical modeling of wet cooling towers–Part 1: Mathematical and physical models." *Journal of Heat Transfer, ASME Transactions,* Vol. 105. pp. 728–735.

Majumdar, A.K.; Singhal, A.K.; Reilly, H.E.; and Bartz, J.A. 1983. "Numerical modeling of wet cooling towers–Part 2: Application to natural and mechanical draft towers." *Journal of Heat Transfer*, ASME Transactions, Vol. 105. pp. 736–743.

McQuiston, F.C., and Parker, J.D. 1982. *Heating, ventilating, and air conditioning*, second edition. New York: John Wiley and Sons.

Stoecker, W.F. 1958. *Refrigeration and air conditioning*. New York: McGraw-Hill Book Company.

Sutherland, J.W. 1983. "Analysis of mechanical-draught counterflow air/water cooling towers." *Journal of Heat Transfer, ASME Transactions*, Vol. 105. pp. 576–583.

Threlkeld, J.L. 1970. *Thermal environmental engineering*, second edition. Englewood Cliffs, New Jersey: Prentice-Hall, Inc.

Villacres, A. 1984. *Computer simulation of performance of evaporative heat exchangers*. M.Sc. Thesis, Pennsylvania State University.

Webb, R.L. 1984. "A unified theoretical treatment for thermal analysis of cooling towers, evaporative condensers, and fluid coolers. *ASHRAE Transactions*, Vol. 90, Part 2. pp. 398–415.

Whiller, A. 1976. "A fresh look at the calculation of performance of cooling towers." *ASHRAE Transactions*, Vol. 82, Part 1. pp. 269–282.

Fundamental Principle Algorithms

Oachs, B.W. 1982. *Computer modeling of evaporative cooling towers*. M.Sc. Thesis, North Dakota State University.

3.7.2 Indirect-Contact Towers

Equipment Description:

An indirect-contact tower is a device whose purpose is to cool water by transferring energy to the atmosphere. It uses the evaporation of water into the air as well as the transfer of sensible heat to the air to achieve its cooling effect. A closed-circuit heat exchanger separates the water being cooled from the cooling air and from the water sprayed into the air to cool it.

Relevant Handbook Chapter:

ASHRAE. 1988. Chapter 20, "Cooling towers." *ASHRAE handbook–1988 equipment*. Atlanta: American Society of Heating, Refrigerating, and Air-Conditioning Engineers, Inc.

Key References:

Semi-Theoretical Algorithms

Webb, R.L., and Villacres, A. 1984. "Algorithms for performance simulation of cooling towers, evaporative condensers, and fluid coolers." *ASHRAE Transactions*, Vol. 90, Part 2. pp. 416–458.

Webb, R.L., and Villacres, A. 1984. "Performance simulation of evaporative heat exchangers (Cooling towers, fluid coolers, and condensers)." *AIChE Symposium Session*, Vol. 79. pp. 458–464.

Semi-Theoretical Models

Cheng, M.S. 1976. *Analysis of different types of dry-wet cooling towers*. Ph.D. Thesis, University of Iowa.

Jalali, E.A. 1979. *Cooling tower analyses and economic evaluation*. M.Sc. Thesis, California State University, Long Beach.

Villacres, A. 1984. *Computer simulation of performance of evaporative heat exchangers*. M.Sc. Thesis, Pennsylvania State University.

Webb, R.L. 1984. "A unified theoretical treatment for thermal analysis of cooling towers, evaporative condensers, and fluid coolers." *ASHRAE Transactions*, Vol. 90, Part 2. pp. 398–415.

4. HEATING EQUIPMENT

4.1 BOILERS

Equipment Description:

A boiler is a pressure vessel designed to transfer heat to liquid water to produce hot water or steam for use in distribution networks. The heat is produced by burning fossil fuels, by electrical resistance elements, or by direct action of electrodes on the water.

Relevant Handbook Chapter:

ASHRAE. 1988. Chapter 23, "Boilers." *ASHRAE handbook - 1988 equipment*. Atlanta: American Society of Heating, Refrigerating and Air-Conditioning Engineers, Inc.

Key References:

Semi-Empirical Algorithms

ASHRAE. 1976. *Energy calculations 2 - Procedures for simulating the performance of components and systems for energy calculations*. Atlanta: American Society of Heating, Refrigerating and Air-Conditioning Engineers, Inc. p. 31.

ASHRAE. 1983. *Simplified energy analysis using the modified bin method*. Atlanta: American Society of Heating, Refrigerating and Air-Conditioning Engineers, Inc. pp. 6-4 - 6-7.

BLAST. 1986. *Building loads analysis and system thermodynamics program: User's manual, Version 3.0*. U.S. Army Construction Engineering Research Laboratory, Champaign, Illinois. pp. 6-27 - 6-31.

Bonne, U., and Langmead, J.P. 1978. "A short method to determine seasonal efficiency of fossil-fired heating systems for labeling purposes." *ASHRAE Transactions*, Vol. 84, Part 1. pp. 422-446.

DOE-2. 1982. *Engineer's manual, Version 2.1A*. Lawrence Berkeley Laboratory, Berkeley, California, and Los Alamos National Laboratory, Los Alamos, New Mexico. NTIS, Reference DE83004575. pp. V.18-V.26.

HAP. 1987. *E20-II–Hourly analysis program: User's manual, Version 1.1, Vol. 4, Energy analysis, Documentation guide*. Carrier Corporation, Syracuse, New York. pp. 6-1 - 6-2.

TrakLoad. 1986. *Energy audit system: Reference manual, Version 3.1*. Morgan Systems Corporation, Berkeley, California. pp. 4-41 - 4-42.

Semi-Empirical Models

Gajanana, B.C. 1979. *Computer simulation and application studies of a space conditioning system with a diesel driven chiller and dual tank thermal storage*. Ph.D. Thesis, Drexel University, Philadelphia, Pennsylvania. pp. 42-43.

TRACE. 1986. "Equipment documentation." *Trane Air Conditioning Economics Program: User's Manual, Version 500*. The Trane Company, La Crosse, Wisconsin. p. 13.

Zweifel, G. 1986. "A simulation model for oil fired condensing boilers." *Proceedings of the Second International Conference on System Simulation in Buildings*. Liege, Belgium.

Semi-Theoretical Algorithms

Dachelet, M.; Eppe, J.P.; Hannay, J.; Laret, L.; Lebrun, J.; Liebecq, G.; and Lorea, B. 1987. *Fuel oil boiler*. International Energy Agency, Annex 10, System Simulation, University of Liege, Belgium.*

Kelly, G.E., Chi, J.; and Kuklewicz, M.E. 1978. *Recommended testing and calculation procedures for determining the seasonal performance of residential central furnaces and boilers*. NBSIR 78-1543. National Bureau of Standards, Washington, D.C.*

Semi-Theoretical Models

Claus, G.; and Stephan, W. 1985. "A general computer simulation model for furnaces and boilers." *ASHRAE Transactions*, Vol. 91, Part 1B. pp. 47-58.*

Lebrun, J.J.; Hannay, J.; Dols, J.M.; and Morant, M.A. 1985. "Research of a good boiler model for HVAC energy simulation. *ASHRAE Transactions*, Vol. 91, Part 1B. pp. 60-86.*

Malmstrom, T.G.; Mundt, B.; and Bring, A.G. 1985. "A simple boiler model." *ASHRAE Transactions*, Vol. 91, Part 1B. pp. 87-108.*

Meijnen, A.J. 1985. *Development of a computer model for the selection of boiler plants from the viewpoint of a low energy consumption*. Contract No. EEA-1-001-N. Final Report of the Commission of the European Communities.*

Fundamental Principle Algorithms

Chi, J.; Chern, L.; and Didion, D.A. 1983. *A commercial heating boiler transient analysis simulation model (DEPAB2)*. NBSIR 83-2638, National Bureau of Standards, Washington, D.C.*

Goldschmidt, I.G. 1979. *Development of a residential oil-fired heating system model*. M.Sc. Thesis, University of Virginia.*

Fundamental Principle Models

Berkowitz, D.A.; Jones, R.T.; and Markunas, A.L. 1974. "Dynamic model for Mystic No. 4." *Proceedings of the Seminar on Boiler Modeling*. The Mitre Corporation, Bedford, Massachusetts.*

Chi, J. 1976. "Computer simulation of fossil-fuel-fired boilers." *Proceedings of the Conference on Improving Efficiency and Performance of HVAC Equipment and Systems for Commercial and Industrial Buildings*. Purdue University. pp. 336-346.*

Chi, J. 1977. "Seasonal operating performance of gas-fired hydronic heating systems with certain energy-saving features." *Proceedings of the Centre for Heating and Mass Transfer*. Dubrovnik, Yugoslavia.*

Clarke, J.A. 1985. *Energy simulation in building design*. Boston: Adam Hilger Ltd. pp. 262-264.*

Dieck-Assad, G. 1983. *Medium scale modeling and controller optimization of a boiler-turbine system*. Ph.D. Thesis, University of Texas at Austin.*

Hanby, V.I., and Clarke, J.A. 1987. *Catalogue of available HVAC plant component models in the U.K.* Report to SERC GR/D/07459. University of Strathclyde, Glasgow. Code A8.*

Laubli, F. 1974. "Large scale boiler simulation." *Proceedings of the Seminar on Boiler Modeling*. The Mitre Corporation, Bedford, Massachusetts.*

Related References:

ASHRAE. 1983. "Methods of testing for heating seasonal efficiency of central furnaces and boilers." *ASHRAE Standard 103-1982*. Atlanta: American Society of Heating, Refrigerating and Air-Conditioning Engineers, Inc.

4.2 FURNACES

Equipment Description:

A furnace is a device designed to be the principal heating source for the occupied spaces of a building. Heat generated in a furnace by combustion of fossil fuels is transferred by conduction through heat exchange surfaces to air that is used in distribution networks. Alternatively, one or more electric resistance elements directly heat the air. The heated air is circulated by natural convection or by a fan or blower.

Relevant Handbook Chapter:

ASHRAE. 1988. Chapter 24, "Furnaces." *ASHRAE handbook - 1988 equipment*. Atlanta: American Society of Heating, Refrigerating and Air-Conditioning Engineers, Inc.

Key References:

Semi-Empirical Algorithms

ASHRAE. 1976. *Energy calculations 2 - Procedures for simulating the performance of components and systems for energy calculations*. Atlanta: American Society of Heating, Refrigerating and Air-Conditioning Engineers, Inc. p. 31.

Bonne, U., and Langmead, J.P. 1978. "A short method to determine seasonal efficiency of fossil-fired heating systems for labeling purposes." *ASHRAE Transactions*, Vol. 84, Part 1. pp. 422-446.

DOE-2. 1982. *Engineer's manual, Version 2.1A*. Lawrence Berkeley Laboratory, Berkeley, California, and Los Alamos National Laboratory, Los Alamos, New Mexico. NTIS, Reference DE83004575. pp. V.22-V.23.

ENERPASS. 1987. *A computer program for the estimation of building energy consumption, Version 2.1*. Enermodal Engineering Ltd., Waterloo, Ontario. p. 86.

HAP. 1987. *E20-II–Hourly analysis program: user's manual, Version 1.1, Vol. 4, Energy analysis, Documentation guide*. Carrier Corporation, Syracuse, New York. pp. 6-1–6-2.

Huizenga, C., and Barnaby, C.S. 1985. "Methodology for hourly simulation of residential HVAC equipment." *Proceedings of the First National Conference on Microcomputer Applications for Conservation and Renewable Energy*, February 26-28. pp. 5-6.

TrakLoad. 1986. *Energy audit system: Reference manual, Version 3.1*. Morgan Systems Corporation, Berkeley, California. pp. 4-41–4-42.

Semi-Empirical Models

Goldschmidt, V.W.; White, R.R.; and Leonard, R.G. 1980. "Field measurement of a gas furnace in a mobile home and determination of steady-state, cyclic and seasonal efficiencies." *ASHRAE Transactions*, Vol. 86, Part 1. pp. 379-393.

Hirsch, J.J. 1982. "Simulation of HVAC equipment in the DOE-2 program." *Proceedings of the First International Conference on System Simulation in Buildings*. Liege, Belgium. pp. 101-102.

TRACE. 1986. "Equipment documentation." *Trane Air Conditioning Economics Program: User's Manual, Version 500*. The Trane Company, La Crosse, Wisconsin. p. 13.

White, R.R., and Goldschmidt, V.W. 1976. "Off-design performance of a gas furnace." *Proceedings of the Conference on Improving Efficiency and Performance of HVAC Equipment and Systems for Commercial and Industrial Buildings*. Purdue University. pp. 347-354.

Semi-Theoretical Algorithms

Chi, J., and Kelly, G.E. 1978. "A method for estimating the seasonal performance of residential gas and oil-fired heating systems." *ASHRAE Transactions*, Vol. 84, Part 1. pp. 405-421.*

Kelly, G.E.; Chi, J.; and Kuklewicz, M.E. 1978. *Recommended testing and calculation procedures for determining the seasonal performance of residential central furnaces and boilers*. NBSIR 78-1543. National Bureau of Standards, Washington, D.C.*

Semi-Theoretical Models

Goldschmidt, V.W.; Leonard, R.G.; and White, R.R. 1979. "Mathematical modeling and prediction of seasonal efficiencies of a gas fired furnace in a mobile home." *ASME Publication 79-WA/HT-29*. New York: American Society of Mechanical Engineers.*

Macriss, R.A.; Elkins, R.H.; and Zawacki, T.S. 1978. "Seasonal performance analysis of central furnace installations." *Proceedings of the Conference on Documentation and Analysis of Improvements in Efficiency and Performance of HVAC Equipment and Systems*. Purdue University. pp. 78-92.

Pedersen, C.O.; McCulley, M.T.; and Nicol, J.L. 1985. "A mechanistic model for warm-air furnaces." *ASHRAE Transactions*, Vol. 91, Part 2. pp. 131-141.*

Related References:

ASHRAE. 1983. "Methods of testing for heating seasonal efficiency of central furnaces and boilers." *ASHRAE Standard 103-1982*. Atlanta: American Society of Heating, Refrigerating and Air-Conditioning Engineers, Inc.

ASHRAE. 1987. "Methods of testing for rating non-residential warm air heaters." *ASHRAE Standard 45-1986*. Atlanta: American Society of Heating, Refrigerating and Air-Conditioning Engineers, Inc.

Gable, G.K., and Koenig, K. 1977. "Seasonal operating performance of gas heating systems with certain energy-saving features." *ASHRAE Transactions*, Vol. 83, Part 2. pp. 850-864.

4.3 UNIT VENTILATORS AND UNIT HEATERS

Equipment Description:

A unit ventilator or a unit heater is an assembly of a fan and motor, a heating element, and an enclosure whose function is to heat a space. Filters, dampers, directional outlets, duct collars, combustion chambers, and flues can also be included. In addition, the unit ventilator, which is applied primarily to classrooms, has an outside air connection and can have a mechanical cooling stage to provide cooling and dehumidification. The heat source for these types of equipment can be a steam coil, a hot-water coil, a fossil-fuel-fired burner, or electric resistance elements.

Relevant Handbook Chapter:

ASHRAE. 1988. Chapter 27, "Unit ventilators, unit heaters, and makeup air units." *ASHRAE handbook - 1988 equipment*. Atlanta: American Society of Heating, Refrigerating and Air-Conditioning Engineers, Inc.

Key References:

Semi-Empirical Algorithms
DOE-2. 1982. *Engineer's manual, Version 2.1A*. Lawrence Berkeley Laboratory, Berkeley, California, and Los Alamos National Laboratory, Los Alamos, New Mexico. NTIS, Reference DE83004575. pp. IV.165-IV.169.

4.4 HEAT DISTRIBUTION EQUIPMENT

4.4.1 Radiators

Equipment Description:

A radiator is a heat-distributing device used in steam and low-temperature-water heating systems. It supplies heat to a room or zone through a combination of radiation and convection. This device generally consists of an assembly of sectional cast-iron columns that is not enclosed.

Relevant Handbook Chapter:

ASHRAE. 1988. Chapter 28, "Radiators, convectors, baseboard, and finned-tube units." *ASHRAE handbook - 1988 equipment*. Atlanta: American Society of Heating, Refrigerating and Air-Conditioning Engineers, Inc.

Key References:

Semi-Empirical Algorithms
Stephan, W. 1988. *Radiator*. International Energy Agency, Annex 10, System Simulation, Report

AN10-880419-04, University of Liege, Belgium.*

Semi-Empirical Models

IEA. 1986. "EMPA model." *System simulation: Synthesis report for three years of activities (1983-1985) - The simulation exercises, Level 1*. Document AN10 860429-01. IEA Annex 10, University of Liege, Belgium. pp. 21-22.

IEA. 1986. "IKE model." *System simulation: Synthesis report for three years of activities (1983-1985) - The simulation exercises, Level 1*. Document AN10 860429-01. IEA Annex 10, University of Liege, Belgium. pp. 22-23.*

IEA. 1986. "RIT model." *System simulation: Synthesis report for three years of activities (1983-1985) - The simulation exercises, Level 1*. Document AN10 860429-01. IEA Annex 10, University of Liege, Belgium. pp. 23-24.*

IEA. 1986. "ULG model." *System simulation: Synthesis report for three years of activities (1983-1985) - The simulation exercises, Level 1*. Document AN10 860429-01. IEA Annex 10, University of Liege, Belgium. p. 24.*

Fundamental Principle Models

Clarke, J.A. 1985. *Energy simulation in building design*. Boston: Adam Hilger Ltd. pp. 264-266.*

Hanby, V.I., and Clarke, J.A. 1987. *Catalogue of available HVAC plant component models in the U.K.* Report to SERC GR/D/07459. University of Strathclyde, Glasgow. Code A3.*

Hanby, V.I., and Clarke, J.A. 1987. *Catalogue of available HVAC plant component models in the U.K.* Report to SERC GR/D/07459. University of Strathclyde, Glasgow. Code A4.*

Maver, T.W., and Clarke, J.A. 1987. *Development of a prototypical component-based energy modelling system*. ABACUS Final Report for Grant GR/D/26610. University of Strathclyde, Glasgow. pp. 24-25.*

Related References:

Morant, M. 1985. *Exercises on the building "La Chaumiere," Simulation of the heating system, Exercises 1-b and 1-c: Simulation of a section of the heat distribution system and simulation of the whole system, Analysis of results*. International Energy Agency, Annex 10, System Simulation, Report AN10-850308-01, University of Liege, Belgium.

4.5 INFRARED HEATING EQUIPMENT

Equipment Description:

Infrared heating equipment includes infrared radiant heaters and panel heating systems, which directly heat objects within a room. Infrared radiant heaters operate at source temperatures from 350° to 5000°F (180° to 2760°C). They are compact, self-contained devices that consist of an electric or fossil-fuel-fired infrared heating section and reflectors that control the radiation distribution pattern. Panel heating systems consist of hot-water pipes or electric resistance elements that are embedded in ceilings, floors, or walls. In some cases, warm air is circulated through cavities behind or under the panels instead. Panel surface temperatures are generally much lower than those of infrared heaters.

Relevant Handbook Chapters:

ASHRAE. 1987. Chapter 7, "Panel heating and cooling systems." *ASHRAE handbook - 1987 systems and applications*. Atlanta: American Society of Heating, Refrigerating and Air-Conditioning Engineers, Inc.

ASHRAE. 1987. Chapter 16, "Infrared radiant heating." *ASHRAE handbook - 1987 systems and applications*. Atlanta: American Society of Heating, Refrigerating and Air-Conditioning Engineers, Inc.

ASHRAE. 1988. Chapter 29, "Infrared heaters." *ASHRAE handbook - 1988 equipment*. Atlanta: American Society of Heating, Refrigerating and Air-Conditioning Engineers, Inc.

Key References:

Semi-Empirical Models
Maloney, D.M.; Pedersen, C.O.; and Witte, M.J. 1988. "Development of a radiant heating system
    model for BLAST." *ASHRAE Transactions*, Vol. 94, Part 1.
Fundamental Principle Models
Arneodo, P.; Mazza, A.; and Oliaro, P. 1988. *Floor heating panels*. International Energy Agency,
    Annex 10, System Simulation, Report AN10-880520-06, University of Liege, Belgium.*

5. UNITARY EQUIPMENT

5.1 ROOM AIR CONDITIONERS AND PACKAGED TERMINAL AIR CONDITIONERS

Equipment Description:

Room air conditioners and packaged terminal air conditioners are mechanical vapor-compression machines that use an air-cooled or water-cooled condenser to reject heat when cooling and/or dehumidifying indoor air from an individual room or zone. These types of air conditioners consist of encased assemblies designed primarily for mounting through a wall or, as in the case of room air conditioners, in a window. Their basic components include an indoor air coil (refrigerant evaporator), a compressor (and its drive), a condenser, and refrigerant flow-control devices. They can also include provisions for air filtering or cleaning, ventilation, and heating. The heat source can be a fossil-fuel-fired burner, electric resistance elements, or hot-water or steam coils.

Relevant Handbook Chapter:

ASHRAE. 1988. Chapter 41, "Room air conditioners and dehumidifiers"; Chapter 43, "Applied packaged equipment." *ASHRAE handbook–1988 equipment*. Atlanta: American Society of Heating, Refrigerating, and Air-Conditioning Engineers, Inc.

Key References:

Semi-Empirical Algorithms
DOE-2. 1982. *Engineer's manual, Version 2.1A*. Lawrence Berkeley Laboratory, Berkeley, California, and Los Alamos National Laboratory, Los Alamos, New Mexico. NTIS, Reference DE83004575. pp. IV.154–IV.164.
TrakLoad. 1986. *Energy audit system: Reference manual, Version 3.1*. Morgan Systems Corporation, Berkeley, California. pp. 4-35–4-42.

Related References:

ASHRAE. 1984. "Methods of testing for rating room air conditioners and packaged terminal air conditioners." *ASHRAE Standard 16–1983*. Atlanta: American Society of Heating, Refrigerating, and Air-Conditioning Engineers, Inc.

5.2 UNITARY AIR CONDITIONERS

Equipment Description:

A unitary air conditioner is a mechanical vapor-compression machine that uses an air-cooled, water-cooled, or evaporation-cooled condenser to reject heat when cooling and/or dehumidifying indoor air. This type of air conditioner consists of one or more factory-matched assemblies. Its basic components include an indoor air coil (refrigerant evaporator), a compressor (and its drive), a condenser, and refrigerant flow-control devices. Fossil-fuel-fired burner, electric resistance element, hot-water coil, or steam coil heating sections can be incorporated with the air conditioner to provide air heating capability. Fans, air filters, and airflow dampers can also be included to circulate air through an attached distribution network.

Relevant Handbook Chapter:

ASHRAE. 1988. Chapter 42, "Unitary air conditioners and unitary heat pumps." *ASHRAE handbook–1988 equipment*. Atlanta: American Society of Heating, Refrigerating, and Air-Conditioning Engineers, Inc.

Key References:

Empirical Models
Herold, K.E.; Fischer, R.D.; Jakob, F.E; Cudnik, R.A.; and Locklin, D.W. 1987. *Dynamic and seasonal performance of central forced-air systems installed in residences*. Summary Report of ASHRAE Special Project SP43: Phase III. Cooling. Battelle, Columbus, OH. p. IV-3.
Semi-Empirical Algorithms
Huizenga, C., and Barnaby, C.S. 1985. "Methodology for hourly simulation of residential HVAC equipment." *Proceedings of the First National Conference on Microcomputer Applications for Conservation and Renewable Energy*, February 26–28.
Kelly, G.E., and Parken, W.H. 1978. *Method of testing, rating, and estimating the seasonal performance of central air conditioners and heat pumps operating in the cooling mode*. Report No. NBSIR-77-1271, National Bureau of Standards, Center for Building Technology, Washington, D.C.
Semi-Empirical Models
Bullock, C.E. 1984. "Dynamic simulation models for commercial air conditioning and heat pump systems." *Proceedings of the Workshop on HVAC Controls Modeling and Simulation*. Atlanta.*
Semi-Theoretical Algorithms
ASHRAE. 1976. *Energy calculations 2–Procedures for simulating the performance of components and systems for energy calculations*. Atlanta: American Society of Heating, Refrigerating, and Air-Conditioning Engineers, Inc. pp. 59–60.
ASHRAE. 1983. *Simplified energy analysis using the modified bin method*. Atlanta: American Society of Heating, Refrigerating, and Air-Conditioning Engineers, Inc. pp. 4-24–4-31.
BLAST. 1986. *Building loads analysis and system thermodynamics program: User's manual, Version 3.0*. U.S. Army Construction Engineering Research Laboratory, Champaign, Illinois. pp. 5-35–5-44.
Hittle, D.C. 1982. "An algorithm for modeling a direct expansion air-cooled condensing unit." *ASHRAE Transactions*, Vol. 88, Part 2. pp. 655–676.
TrakLoad. 1986. *Energy audit system: Reference manual, Version 3.1*. Morgan Systems Corporation, Berkeley, California. pp. 4-35–4-38.

Related References:

ASHRAE. 1978. "Methods of testing for rating unitary air conditioning and heat pump equipment." *ASHRAE Standard 37–1978*. Atlanta: American Society of Heating, Refrigerating, and Air-Conditioning Engineers, Inc.
ASHRAE. 1980. "Methods of testing for rating positive displacement condensing units." *ASHRAE Standard 14–80*. Atlanta: American Society of Heating, Refrigerating, and Air-Conditioning Engineers, Inc.
ASHRAE. 1983. "Methods of testing for seasonal efficiency of unitary air-conditioners and heat pumps." *ASHRAE Standard 116–1983*. Atlanta: American Society of Heating, Refrigerating, and Air-Conditioning Engineers, Inc.
Chalifoux, A.T. 1986. *A generalized means of simulating HVAC distribution systems on a computer*. M.Sc. Thesis, University of Illinois at Urbana–Champaign.

5.3 UNITARY HEAT PUMPS

5.3.1 Air-to-Air Unitary Heat Pumps

Equipment Description:

An air-to-air unitary heat pump is a mechanical vapor-compression machine that uses an outdoor coil (refrigerant-to-air heat exchanger) to reject heat to outdoor air when cooling and/or dehumidifying indoor air, and to extract heat from outdoor air when heating indoor air. This type of heat pump consists of one or more factory-matched assemblies. Its basic components include an indoor air coil, a compressor (and its drive), an outdoor air coil, a refrigerant or airflow-directing valve, and other refrigerant flow-control devices. Supplementary heaters can be included as part of these heat pumps to provide heat during defrost cycles or during periods of high heat demand that cannot be satisfied by the heat pump alone.

Relevant Handbook Chapter:

ASHRAE. 1988. Chapter 42, "Unitary air conditioners and unitary heat pumps." *ASHRAE handbook–1988 equipment*. Atlanta: American Society of Heating, Refrigerating, and Air-Conditioning Engineers, Inc.

Key References:

Empirical Models
Crawford, R.R., and Shirey, D.B. 1987. "Dynamic modeling of a residential heat pump from actual system performance data." *ASHRAE Transactions*, Vol. 93, Part 2.*
Goldschmidt, V.W., and Hart, G.H. 1982. "Heat pump system performance: Experimental and theoretical results." *ASHRAE Transactions*, Vol. 88, Part 1. pp. 479–489.
Semi-Empirical Algorithms
Carroll, J.T. 1984. *Computer simulation of heat pump systems in residential applications*. M.Sc. Thesis, Department of Mechanical and Aerospace Engineering, North Carolina State University.
HAP. 1987. *E20-II–Hourly analysis program: User's manual, Version 1.1, Vol. 4, Energy analysis, Documentation guide*. Carrier Corporation, Syracuse, New York. pp. 7-1-7-2.
Howe, R.R. 1983. *Model and performance characteristics of a commercially-sized hybrid air conditioning system which utilizes a rotary desiccant dehumidifier*. M.Sc. Thesis, University of Wisconsin–Madison. pp. 25–27.
Huizenga, C., and Barnaby, C.S. 1985. "Methodology for hourly simulation of residential HVAC equipment." *Proceedings of the First National Conference on Microcomputer Applications for Conservation and Renewable Energy*, February 26–28. pp. 1–4.
Kleiser, J.D. 1987. *Modeling internal combustion engine-powered desiccant cooling systems*. M.Sc. Thesis, University of Nevada, Reno. pp. 22–23.
Parken, W.H.; Kelly, G.E.; and Didion, D.A. 1980. *Method of testing, rating, and estimating the heating seasonal performance of heat pumps*. Report No. NBSIR-80-2002, National Bureau of Standards, National Engineering Lab, Washington, D.C.
TRACE. 1986. "Equipment documentation." *Trane Air Conditioning Economics Program: User's Manual, Version 500*. The Trane Company, La Crosse, Wisconsin. pp. 13–14.
TrakLoad. 1986. *Energy audit system: Reference manual, Version 3.1*. Morgan Systems Corporation, Berkeley, California. pp. 4-35–4-39.
Semi-Empirical Models
Bonne, U., and Patani, A. 1978. "Modeling the influence of heat pump sizing, climate, and test conditions on seasonal efficiency (COP)." *Proceedings of the Conference on Documentation and Analysis of Improvements in Efficiency and Performance of HVAC Equipment and Systems*, Purdue University. pp. 71–77.
Bonne, U.; Jacobson, R.D.; Patani, A.; and Aagard, R.L. 1978. "Seasonal efficiency of residential

41

combustion and heat pump systems." *Proceedings of the International Symposium on Simulation, Modeling and Decision in Energy Systems*, Montreal, Canada: The International Association of Science and Technology for Development (IASTED). pp. 277–287.

Bullock, C.E. 1984. "Dynamic simulation models for commercial air conditioning and heat pump systems." *Proceedings of the Workshop on HVAC Controls Modeling and Simulation*. Atlanta.*

Chen, R.F. 1988. *Assessments of desiccant cooling systems*. M.Sc. Thesis, University of Nevada, Reno.

EMPS-2.1. 1988. *Computer program for residential building energy analysis–Engineering manual*. Electric Power Research Institute, Palo Alto, California. pp. 3-8–3-12.

Groff, G.C., and Bullock, C.E. 1976. "A computer simulation model for air-source heat pump system seasonal performance studies." *Proceedings of the Second Annual Heat Pump Technology Conference*. Oklahoma State University, Stillwater, Oklahoma.*

Hart, G.H., and Goldschmidt, V.W. 1980. "Field measurements of a mobile home unitary heat pump (in a heating mode)." *ASHRAE Transactions*, Vol. 86, Part 2. pp. 347–367.*

Hirsch, J.J. 1982. "Simulation of HVAC equipment in the DOE-2 program." *Proceedings of the First International Conference on System Simulation in Buildings*. Liege, Belgium. p. 102.

Rice, C.K.; Fischer, S.K.; and Emerson, C.J. 1986. *The Oak Ridge heat pump models: II. An annual performance factor/loads model for residential air-source heat pumps*. ORNL/CON-160, Oak Ridge National Laboratory.

TRNSYS. 1981. *A transient system simulation program: Reference manual*. Solar Energy Laboratory, University of Wisconsin–Madison. pp. 4.20-1–4.20-12.

Turner, R.H.; Kleiser, J.D.; Chen, R.F.; Domingo, N.; and Chen, F. 1988. "Assessment of thermally activated heat pumps with desiccant cooling." *ASHRAE Transactions*, Vol. 94, Part 1.

Fundamental Principle Algorithms

Olivier, P.P.; Sepsy, C.F.; and Jones, C.D. 1973. *A thermodynamic computer simulation of an air-to-air heat pump system for a commercial building*. Progress Report, Department of Mechanical Engineering, Ohio State University, Columbus, Ohio.

Fundamental Principle Models

Fischer, S.K., and Rice, C.K. 1983. *The Oak Ridge heat pump models: I. A steady-state computer design model for air-to-air heat pumps*. ORNL/CON-80/R1, Oak Ridge National Laboratory.

Related References:

ASHRAE. 1978. "Methods of testing for rating unitary air conditioning and heat pump equipment." *ASHRAE Standard 37-1978*. Atlanta: American Society of Heating, Refrigerating, and Air-Conditioning Engineers, Inc.

ASHRAE. 1983. "Methods of testing for seasonal efficiency of unitary air-conditioners and heat pumps." *ASHRAE Standard 116-1983*. Atlanta: American Society of Heating, Refrigerating, and Air-Conditioning Engineers, Inc.

Botkin, H.C.; Dabiri, A.E.; and Blatt, M.H. 1983. "Heat pump performance parametric studies using seasonal performance model (SPM)." *ASHRAE Transactions*, Vol. 89, Part 1B. pp. 548–556.

Parken, W.H.; Beausoliel, R.W.; and Kelly, G.E. 1977. "Factors affecting the performance of a residential air-to-air heat pump." *ASHRAE Transactions*, Vol. 83, Part 2. pp. 839–849.

Wildin, M.W.; Fong, A.; Nakos, J.; and Nakos, J. 1978. "Analysis of performance of installed air-to-air heat pumps using a simple method for predicting the effects of cycling on heating performance in an arid climate." *Proceedings of the Conference on Documentation and Analysis of Improvements in Efficiency and Performance of HVAC Equipment and Systems*. Purdue University. pp. 60–70.

5.3.2 Air-to-Water Unitary Heat Pumps

Equipment Description:

An air-to-water unitary heat pump is a mechanical vapor-compression machine that uses a refrigerant-to-liquid heat exchanger to reject heat to a liquid when cooling and/or dehumidifying indoor air, and to extract heat from a liquid when heating indoor air. The liquid can be water, brine, or ethylene-glycol solution. This type of heat pump consists of one or more factory-matched assemblies. Its basic components include an indoor air coil, a compressor (and its drive), a refrigerant-to-liquid heat exchanger, a refrigerant flow-directing valve, and other refrigerant flow-control devices.

Relevant Handbook Chapter:

ASHRAE. 1988. Chapter 43, "Applied packaged equipment." *ASHRAE handbook–1988 Equipment*. Atlanta: American Society of Heating, Refrigerating, and Air-Conditioning Engineers, Inc.

Key References:

Empirical Algorithms
ASHRAE. 1976. *Energy calculations 2–Procedures for simulating the performance of components and systems for energy calculations*. Atlanta: American Society of Heating, Refrigerating, and Air-Conditioning Engineers, Inc. pp. 90–93.
Semi-Empirical Algorithms
Carroll, J.T. 1984. *Computer simulation of heat pump systems in residential applications*. M.Sc. Thesis, Department of Mechanical and Aerospace Engineering, North Carolina State University.
DOE-2. 1982. *Engineer's manual, Version 2.1A*. Lawrence Berkeley Laboratory, Berkeley, California, and Los Alamos National Laboratory, Los Alamos, New Mexico. NTIS, Reference DE83004575. pp. IV.143–IV.151.
HAP. 1987. *E20-II–Hourly analysis program: User's manual, Version 1.1, Vol. 4, Energy analysis, Documentation guide*. Carrier Corporation, Syracuse, New York. pp. 7-2-7-6.
TrakLoad. 1986. *Energy audit system: Reference manual, Version 3.1*. Morgan Systems Corporation, Berkeley, California. pp. 4-35-4-40.
Semi-Empirical Models
Conde, M.R. 1986. *Technical form: Air-to-water heat pump simple simulation model*. International Energy Agency, Annex 10, System Simulation, Report AN10-861022-06, University of Liege, Belgium.
TRNSYS. 1981. *A transient system simulation program: Reference manual*. Solar Energy Laboratory, University of Wisconsin–Madison. pp. 4.20-1–4.20-12.
Semi-Theoretical Models
EMPS-2.1. 1988. *Computer program for residential building energy analysis–Engineering manual*. Electric Power Research Institute, Palo Alto, California. pp. 3-8-3-19.

Related References:

ASHRAE. 1978. "Methods of testing for rating unitary air conditioning and heat pump equipment." *ASHRAE Standard 37–1978*. Atlanta: American Society of Heating, Refrigerating, and Air-Conditioning Engineers, Inc.
Chimack, M.J. 1987. *Development of a heat pump simulation model for the BLAST energy analysis computer program*. M.Sc. Thesis, University of Illinois at Urbana–Champaign.

6. SOLAR HEATING EQUIPMENT

6.1 SOLAR COLLECTORS

Equipment Description:

A solar collector is a heat exchanger that heats a gas or a liquid under forced circulation by absorbing direct and/or diffuse incident solar radiation. The collector can be a flat-plate type or a concentrating type. The absorbing surface in the flat-plate type is essentially planar, with similar areas and geometry for both the aperture and absorber. In the concentrating collector, reflectors, lenses, or other optical elements redirect and/or concentrate the solar radiation passing through the aperture to the absorber, which can have a smaller area than the aperture.

Relevant Handbook Chapter:

ASHRAE. 1988. Chapter 44, "Solar energy equipment." *ASHRAE handbook-1988 equipment.* Atlanta: American Society of Heating, Refrigerating and Air-Conditioning Engineers, Inc.

Key References:

Semi-Empirical Algorithms
ASHRAE. 1983. *Solar collector performance manual, ASHRAE SP 32.* Atlanta: American Society of Heating, Refrigerating and Air-Conditioning Engineers, Inc.
BLAST. 1986. *Building loads analysis and system thermodynamics program: User's manual, Version 3.0.* U.S. Army Construction Engineering Research Laboratory, Champaign, Illinois. pp. 6-48-6-50.
DOE-2. 1982. *Engineer's manual, Version 2.1A.* Lawrence Berkeley Laboratory, Berkeley, California, and Los Alamos National Laboratory, Los Alamos, New Mexico. NTIS, Reference DE83004575. pp. V.184-V.204.
Semi-Theoretical Algorithms
Hill, J.E., and Fanney, A.H. 1980. "A proposed procedure of testing for rating solar domestic hot water systems." *ASHRAE Transactions*, Vol. 86, Part 1. pp. 805-822.
Swanson, S.R., and Boehm, R.F. 1977. "Calculation of long term solar collector heating system performance." *Solar Energy*, Vol. 19. pp. 129-138.
Semi-Theoretical Models
ASHRAE. 1986. "Methods of testing to determine the thermal performance of solar collectors." *ASHRAE Standard 93-1986.* Atlanta: American Society of Heating, Refrigerating and Air-Conditioning Engineers, Inc.*
ASHRAE. 1987. "Methods of testing to determine the thermal performance of unglazed flat-plate liquid-type solar collectors." *ASHRAE Standard 96-1987.* Atlanta: American Society of Heating, Refrigerating and Air-Conditioning Engineers, Inc.
Chandrashekar, M.; Le, N.T.; Hollands, K.G.T.; and Sullivan, H.F. 1978. "Simulation study for solar-assisted heat pump systems in Canada." *Proceedings of the International Symposium on*

Simulation, Modeling and Decision in Energy Systems. Montreal, Canada: The International Association of Science and Technology for Development (IASTED). pp. 180–183.

Diab, T.A. 1984. *The analysis of thermally stratified storage tanks using solar energy systems*. M.Sc. Thesis, Worcester Polytechnic Institute.*

Hawken, P., and Karaki, S. 1984. "Testing and computer simulation of an air based, solar assisted heat pump system." *Proceedings of the ASME Solar Energy Division 6th Annual Conference, Las Vegas, Nevada.*

Hill, J.E., and Streed, E.R. 1976. "A method of testing for rating solar collectors based on thermal performance." *Solar Energy*, Vol. 18. pp. 421–429.

Oonk, R.L.; Beckman, W.A.; and Duffie, J.A. 1975. "Modeling of the CSU heating/cooling system." *Solar Energy*, Vol. 17. pp. 21–28.

Threlkeld, J.L. 1970. *Thermal environmental engineering*, second edition. Englewood Cliffs, New Jersey: Prentice-Hall, Inc. pp. 424–434.

Fundamental Principle Models

TRNSYS. 1981. "Flat-plate solar collector." *A transient system simulation program: Reference manual*. Solar Energy Laboratory, University of Wisconsin-Madison. pp. 4.1-1–4.1-20.

TRNSYS. 1981. "CPC collector." *A Transient System Simulation Program: Reference Manual*. Solar Energy Laboratory, University of Wisconsin-Madison. pp. 4.30-1–4.1-7.

Related References

ASHRAE. 1981. "Methods of testing to determine the thermal performance of solar domestic water heating systems." *ASHRAE Standard 95-1981*. Atlanta: American Society of Heating, Refrigerating and Air-Conditioning Engineers, Inc.

Hittle, D.C.; Walton, G.N.; Holshouser, D.F.; and Leverenz, D.J. 1977. *Predicting the performance of solar energy systems*. Report CERL-IR-E-98. Construction Engineering Research Laboratory, Champaign, Illinois.

McQuiston, F.C., and Parker, J.D. 1982. *Heating, ventilating, and air conditioning*, second edition. New York: John Wiley and Sons.

Siebers, D.L., and Viskanta, R. 1977. "Comparison of long-term flat plate solar collector performance calculations based on averaged meteorological data." *Solar Energy*, Vol. 19. pp. 163–169.

Siebers, D.L., and Viskanta, R. 1977. "Comparison of predicted performance of constant outlet temperature and constant mass flow rate collectors." *Solar Energy*, Vol. 19. pp. 411–413.

7. GENERAL

7.1 CENTRIFUGAL PUMPS

Equipment Description:

   A centrifugal pump is a device that creates a pressure difference and causes a liquid to flow by using a power-driven rotating impeller that imparts a centrifugal force on the liquid.

Relevant Handbook Chapter:

ASHRAE. 1988. Chapter 30, "Centrifugal pumps." *ASHRAE handbook–1988 equipment*. Atlanta: American Society of Heating, Refrigerating, and Air-Conditioning Engineers, Inc.

Key References:

Semi-Empirical Algorithms
ASHRAE. 1976. *Energy calculations 2–Procedures for simulating the performance of components and systems for energy calculations*. Atlanta: American Society of Heating, Refrigerating, and Air-Conditioning Engineers, Inc. p. 56.
ASHRAE. 1983. *Simplified energy analysis using the modified bin method*. Atlanta: American Society of Heating, Refrigerating, and Air-Conditioning Engineers, Inc. pp. 6-7–6-8.
BLAST. 1986. *Building loads analysis and system thermodynamics program: User's manual, Version 3.0*. U.S. Army Construction Engineering Research Laboratory, Champaign, Illinois. p. 6–51.
HAP. 1987. *E20-II–Hourly analysis program: User's manual, Version 1.1, Vol. 4, Energy analysis, Documentation guide*. Carrier Corporation, Syracuse, New York. pp. 8-4–8-6.
Silver, S.C.; Jones, J.W.; Peterson, J.L.; Milbitz, A.; and Hunn, B.D. 1988. *CBS/ICE program user's guide*. Center for Energy Studies, The University of Texas at Austin, Conservation and Solar Research Report No. 7, June. pp. 45–48.
TRACE. 1986. "Equipment documentation." *Trane air conditioning economics program: User's manual, Version 500*. The Trane Company, La Crosse, Wisconsin. pp. 11–12.
TrakLoad. 1986. *Energy audit system: Reference manual, Version 3.1*. Morgan Systems Corporation, Berkeley, California. p. 4–35.
Semi-Empirical Models
Clark, D.R. 1985. *HVACSIM+ building systems and equipment simulation program: Reference manual*. NBSIR 84-2996, National Bureau of Standards, Washington, D.C. pp. 31–32.
Gluck, R., and Pollak, E. 1978. "Design optimization of air-conditioning systems." *ASHRAE Transactions*, Vol. 84, Part 2. pp. 304–314.
Marcev, C.L.; Smith, C.R.; and Bruns, D.D. 1984. "Supervisory control and system optimization of chiller-cooling tower combinations via simulation using primary equipment component models: Part I. Primary equipment component models." *Proceedings of the Workshop on HVAC Control Modeling and Simulation*. Atlanta.
Stoecker, W.F. 1980. *Design of thermal systems*, second edition. Toronto: McGraw-Hill Book Company. p. 97.

Semi-Theoretical Algorithms

Brossa, M., and Mazza, A. 1988. *Pump*. International Energy Agency, Annex 10, System Simulation, Report AN10-880520-04, University of Liege, Belgium.

Related References:

Hanby, V.I., and Clarke, J.A. 1987. *Catalogue of available HVAC plant component models in the U.K.* Report to SERC GR/D/07459. University of Strathclyde, Glasgow. Code A9.

McQuiston, F.C., and Parker, J.D. 1982. *Heating, ventilating, and air conditioning.* Second Edition. New York: John Wiley and Sons.

TRNSYS. 1981. *A transient system simulation program: Reference manual.* Solar Energy Laboratory, University of Wisconsin–Madison.

7.2 ENGINES

Equipment Description:

An engine is a device with a rotating shaft that is powered by converting energy developed in the combustion of fossil fuels to mechanical energy. The combustion process increases the pressure of a gas on one or more pistons or rotors, so that a torque is exerted on the shaft. No torque is produced by passing the products of combustion or gas through a rotor or piston.

Relevant Handbook Chapter:

ASHRAE. 1988. Chapter 32, "Engine and turbine drives." *ASHRAE handbook–1988 equipment.* Atlanta: American Society of Heating, Refrigerating, and Air-Conditioning Engineers, Inc.

Key References:

Semi-Empirical Algorithms

ASHRAE. 1976. *Energy calculations 2–Procedures for simulating the performance of components and systems for energy calculations.* Atlanta: American Society of Heating, Refrigerating, and Air-Conditioning Engineers, Inc. p. 61.

BLAST. 1986. *Building loads analysis and system thermodynamics program: User's manual, Version 3.0.* U.S. Army Construction Engineering Research Laboratory, Champaign, Illinois. pp. 6-17–6-21.

Semi-Empirical Models

DOE-2. 1982. *Engineer's manual, Version 2.1A.* Lawrence Berkeley Laboratory, Berkeley, California, and Los Alamos National Laboratory, Los Alamos, New Mexico. NTIS, Reference DE83004575. pp. V.88–V.90.

Gajanana, B.C. 1979. *Computer simulation and application studies of a space conditioning system with a diesel driven chiller and dual tank thermal storage.* Ph.D. Thesis, Drexel University, Michigan.

Kleiser, J.D. 1987. *Modeling internal combustion engine-powered desiccant cooling systems.* M.Sc. Thesis, University of Nevada, Reno.

7.3 TURBINES

7.3.1 Steam Turbines

Equipment Description:

A steam turbine is a device with a rotating shaft that is powered by injecting pressurized steam through nozzles so that the steam can expand as it passes through rows of blades on one or more rotor discs. The turbine can be an axial-flow, a radial-flow, or a mixed-flow type.

Relevant Handbook Chapter:

ASHRAE. 1988. Chapter 32, "Engine and turbine drives." *ASHRAE handbook–1988 Equipment*. Atlanta: American Society of Heating, Refrigerating, and Air-Conditioning Engineers, Inc.

Key References:

Semi-Theoretical Models
Dieck-Assad, G. 1983. *Medium scale modeling and controller optimization of a boiler-turbine system*. Ph.D. Thesis, University of Texas at Austin.*
Jalali, E.A. 1979. *Cooling tower analyses and economic evaluation*. M.Sc. Thesis, California State University, Long Beach.

7.3.2 Gas Turbines

Equipment Description:

A gas turbine is a device with a rotating shaft that is powered by injecting high-pressure combustion products of fossil fuels through nozzles so that they can expand as they pass through rows of blades on one or more rotor discs. The turbine can be an axial-flow, a radial-flow, or a mixed-flow type. Gas turbine components include a compressor and a combustion chamber in addition to the turbine.

Relevant Handbook Chapter:

ASHRAE. 1988. Chapter 32, "Engine and turbine drives." *ASHRAE handbook–1988 equipment* Atlanta: American Society of Heating, Refrigerating, and Air-Conditioning Engineers, Inc.

Key References:

Semi-Empirical Algorithms
BLAST. 1986. *Building loads analysis and system thermodynamics program: User's manual, Version 3.0*. U.S. Army Construction Engineering Research Laboratory, Champaign, Illinois. pp. 6-21–6-25.
DOE-2. 1982. *Engineer's manual, Version 2.1A*. Lawrence Berkeley Laboratory, Berkeley, California, and Los Alamos National Laboratory, Los Alamos, New Mexico. NTIS, Reference DE83004575. pp. V.90–V.91.
Semi-Theoretical Models
Stoecker, W.F. 1980. Design of thermal systems, second edition. Toronto: McGraw-Hill Book Company. pp. 119–122.

7.4 AIR-TO-AIR ENERGY-RECOVERY EQUIPMENT

Equipment Description:

An air-to-air energy-recovery device transfers sensible and/or latent heat from one airstream to another without the addition of heat energy from an external source other than the input from mechanical drives. The energy-recovery device can be one of the following types: a rotary air-to-air energy exchanger, a coil energy-recovery loop, a twin-tower enthalpy-recovery loop, a heat-pipe heat exchanger, a fixed-plate heat exchanger, or a thermosiphon heat exchanger.

Relevant Handbook Chapter:

ASHRAE. 1988. Chapter 34, "Air-to-air energy-recovery equipment." *ASHRAE handbook–1988 equipment*. Atlanta: American Society of Heating, Refrigerating, and Air-Conditioning Engineers, Inc.

Key References:

Semi-Theoretical Algorithms
Barlow, R.S. 1983. *User's manual and report on DESSIM Version 2.0: A Fortran program for simulation of desiccant cooling systems and components*. Mechanical Engineering Department, Stanford University.
Gawley, H.N., and Fisher, D.R. 1975. "The effectiveness and rating of air-to-air heat exchangers." *ASHRAE Transactions*, Vol. 81, Part 2. pp. 401–409.
Kleiser, J.D. 1987. *Modeling internal combustion engine-powered desiccant cooling systems*. M.Sc. Thesis, University of Nevada, Reno.
Semi-Theoretical Models
TrakLoad. 1986. *Energy audit system: Reference manual, Version 3.1*. Morgan Systems Corporation, Berkeley, California. pp. 3-65–3-66.
Fundamental Principle Models
Dastmalchi, B. 1982. *Seasonal simulation of a solar desiccant cooling system*. M.Sc. Thesis, Department of Mechanical Engineering, Colorado State University. pp. 12–20.
Jurinak, J.J. 1982. *Open cycle solid desiccant cooling–Component models and system simulations*. Ph.D. Thesis, University of Wisconsin–Madison.*
Kohonen, R., and Nyman, M. 1987. *Heat recovery devices (HRD)*. International Energy Agency, Annex 10, System Simulation, Report AN10-871117-02, University of Liege, Belgium.*
Pescod, D. 1979. "A heat exchanger for energy saving in an air-conditioning plant." *ASHRAE Transactions*, Vol. 85, Part 2. pp. 238–251.

Related References:

ASHRAE. 1984. "Method of testing air-to-air heat exchangers." *ASHRAE Standard 84–1978*. Atlanta: American Society of Heating, Refrigerating, and Air-Conditioning Engineers, Inc.
Chalifoux, A.T. 1986. *A generalized means of simulating HVAC distribution systems on a computer*. M.Sc. Thesis, University of Illinois at Urbana–Champaign.
Sauer, H.J., and Howell, R.H. 1981. "Promise and potential of air-to-air energy recovery systems." *ASHRAE Transactions*, Vol. 87, Part 1. pp. 167–182.

7.5 THERMAL STORAGE

7.5.1 Sensible Heat Storage Devices

7.5.1.1 Water tanks

Equipment Description:

Water tanks are sensible heat storage devices that use water or a water-glycol solution as a storage medium. Heat absorbed by or removed from the water tank results in an increase or decrease in the temperature of the storage medium without changing the phase of any portion of the storage medium. The device includes the storage medium; its containing structure; and its accessories, such as heat transfer fluids, heat exchangers, flow-switching devices, valves, and baffles.

Relevant Handbook Chapter:

ASHRAE. 1987. Chapter 46, "Thermal storage." *ASHRAE handbook–1987 HVAC systems and applications*. Atlanta: American Society of Heating, Refrigerating, and Air-Conditioning Engineers, Inc.

Key References:

Semi-Theoretical Algorithms
Carroll, J.T. 1984. *Computer simulation of heat pump systems in residential applications*. M.Sc. Thesis, Dept. of Mechanical and Aerospace Engineering, North Carolina State University.*
Fundamental Principle Models
Clarke, J.A. 1985. *Energy simulation in building design*. Boston: Adam Hilger Ltd. pp. 255–256.*
Diab, T.A. 1984. *The analysis of thermally stratified storage tanks using solar energy systems*. M.Sc. Thesis, Worcester Polytechnic Institute.*
Gajanana, B.C. 1979. *Computer simulation and application studies of a space conditioning system with a diesel driven chiller and dual tank thermal storage*. Ph.D. Thesis, Drexel University, Michigan. pp. 32–38.*
Oonk, R.L.; Beckman, W.A.; and Duffie, J.A. 1975. "Modeling of the CSU heating/cooling system." *Solar Energy*, Vol. 17. pp. 21–28.*
TRNSYS. 1981. *A transient system simulation program: Reference manual*. Solar Energy Laboratory, University of Wisconsin–Madison. pp. 4.4-1–4.4-9.*

Related References:

ASHRAE. 1977. "Methods of testing thermal storage devices based on thermal performance." *ASHRAE Standard 94-77*. Atlanta: American Society of Heating, Refrigerating, and Air-Conditioning Engineers, Inc.
ASHRAE. 1986. "Method of testing active sensible thermal energy storage devices based on thermal performance." *ASHRAE Standard 94.3-1986*. Atlanta: American Society of Heating, Refrigerating, and Air-Conditioning Engineers, Inc.
Swanson, S.R., and Boehm, R.F. 1977. "Calculation of long term solar collector heating system performance." *Solar Energy*, Vol. 19. pp. 129–138.

7.5.1.2 Packed rock beds

Equipment Description:

Packed rock beds are sensible heat storage devices that use loosely packed particles (pebbles or rocks) as a storage medium. Heat absorbed by or removed from the packed rock bed results in an increase or decrease in the temperature of the storage medium without changing the phase of any portion of the storage medium. The device includes the storage medium; its containing structure; and its accessories, such as heat transfer fluids, heat exchangers, flow-switching devices, valves, and baffles.

Relevant Handbook Chapter:

ASHRAE. 1987. Chapter 46, "Thermal storage." *ASHRAE handbook–1987 HVAC systems and applications*. Atlanta: American Society of Heating, Refrigerating, and Air-Conditioning Engineers, Inc.

Key References:

Fundamental Principle Models
Clarke, J.A. 1985. *Energy simulation in building design*. Boston: Adam Hilger Ltd. pp. 254–255.*
Hanby, V.I., and Clarke, J.A. 1987. *Catalogue of available HVAC plant component models in the U.K.* Report to SERC GR/D/07459. University of Strathclyde, Glasgow. Code A7.*
TRNSYS. 1981. *A transient system simulation program: Reference manual*. Solar Energy Laboratory, University of Wisconsin–Madison. pp. 4.10-1–4.10-7.*

Related References:

ASHRAE. 1977. "Methods of testing thermal storage devices based on thermal performance." *ASHRAE Standard 94–77*. Atlanta: American Society of Heating, Refrigerating, and Air-Conditioning Engineers, Inc.

ASHRAE. 1986. "Method of testing active sensible thermal energy storage devices based on thermal performance." *ASHRAE Standard 94.3–1986*. Atlanta: American Society of Heating, Refrigerating, and Air-Conditioning Engineers, Inc.

7.5.2 Latent Heat Storage Devices

7.5.2.1 Ice storage devices

Equipment Description:

An ice storage device is a latent heat storage device that stores energy in the latent heat of fusion of ice. Most of the heat added to or removed from the device goes into changing the enthalpy of the ice during a change of phase process. Some heat is stored as sensible heat, since charging and discharging of the device involves a finite change in temperature of the device. The device can be one of the following types: water/ice storage on a refrigerant coil; water/ice storage on a brine coil; a brine-type solid ice builder; a plate ice maker; or an ice slurry maker. These devices include the storage medium (water/ice); a storage tank for water, ice, or ice slurry; coils, tubes, or plates through which refrigerant or brine is circulated; and accessories, such as flow-switching devices, valves, and baffles.

Relevant Handbook Chapter:

ASHRAE. 1987. Chapter 46, "Thermal storage." *ASHRAE handbook–1987 HVAC systems and applications*. Atlanta: American Society of Heating, Refrigerating, and Air-Conditioning Engineers, Inc.

Key References:

Fundamental Principle Algorithms
Silver, S.C.; Jones, J.W.; Peterson, J.L.; Milbitz, A.; and Hunn, B.D. 1988. *CBS/ICE program user's guide*. Center for Energy Studies, The University of Texas at Austin, Conservation and Solar Research Report No. 7, June.*
Fundamental Principle Models
Clarke, J.A. 1985. *Energy simulation in building design*. Boston: Adam Hilger Ltd. p. 256.*
Silver, S.C.; Jones, J.W.; Peterson, J.L.; Milbitz, A.M.; and Hunn, B.D. 1988. "Component models for simulation of ice storage systems." *ASHRAE RP–459 Technical Paper*, presented at ASHRAE Winter Meeting in Dallas.*

Related References:

ASHRAE. 1977. "Methods of testing thermal storage devices based on thermal performance." *ASHRAE Standard 94–77*. Atlanta: American Society of Heating, Refrigerating, and Air-Conditioning Engineers, Inc.

ASHRAE. 1985. "Method of testing active latent heat storage devices based on thermal performance." *ASHRAE Standard 94.1–1985*. Atlanta: American Society of Heating, Refrigerating, and Air-Conditioning Engineers, Inc.

7.5.2.2 Phase change material (PCM) devices

Equipment Description:

A phase change material (PCM) device is a latent heat storage device that stores energy by changing the physical state of a storage medium (other than water/ice), usually from a liquid to a solid (heat of fusion) or vice versa. Solid/solid or liquid/gas transformations can also be used. Most of the heat added to or removed from the device goes into changing the enthalpy of the PCM during a change of phase process. Some heat is stored as sensible heat since charging and discharging of the device involves a finite change in temperature of the device. Typical storage media are salt hydrates, eutectics, organics, and clathrates. The device includes the storage medium; its containing structure; and its accessories, such as heat transfer fluids, heat exchangers, flow-switching devices, valves, and baffles.

Relevant Handbook Chapter:

ASHRAE. 1987. Chapter 46, "Thermal storage." *ASHRAE handbook–1987 HVAC systems and applications*. Atlanta: American Society of Heating, Refrigerating, and Air-Conditioning Engineers, Inc.

Key References:

Fundamental Principle Models
Clarke, J.A. 1985. *Energy simulation in building design*. Boston: Adam Hilger Ltd. p. 256.*

Related References:

ASHRAE. 1977. "Methods of testing thermal storage devices based on thermal performance." *ASHRAE Standard 94–77*. Atlanta: American Society of Heating, Refrigerating, and Air-Conditioning Engineers, Inc.
ASHRAE. 1985. "Method of testing active latent heat storage devices based on thermal performance." *ASHRAE Standard 94.1–1985*. Atlanta: American Society of Heating, Refrigerating, and Air-Conditioning Engineers, Inc.

7.5.3 Electrical Input/Thermal Output Heat Storage Devices

Equipment Description:

Electrical input/thermal output heat storage devices are charged electrically and discharged thermally. Electric resistance elements are used to heat the storage device. A heat transfer fluid flowing through the storage device is used to remove heat from the storage device. The device can store energy as sensible and/or latent heat. Components of the device include electric resistance elements; the storage medium; its containing structure; and its accessories, such as heat transfer fluids, heat exchangers, flow-switching devices, valves, and baffles.

Relevant Handbook Chapter:

ASHRAE. 1987. Chapter 46, "Thermal storage." *ASHRAE handbook–1987 HVAC systems and applications*. Atlanta: American Society of Heating, Refrigerating, and Air-Conditioning Engineers, Inc.

Related References:

ASHRAE. 1981. "Methods of testing thermal storage devices with electrical input and thermal output based on thermal performance." *ASHRAE Standard 94.2–1981*. Atlanta: American Society of Heating, Refrigerating, and Air-Conditioning Engineers, Inc.

7.6 GENERAL HEAT EXCHANGERS

Equipment Description:

A heat exchanger is a device that transfers heat between two fluids. One or both of the fluids is under forced circulation. The term "general heat exchangers" includes all heat exchangers, except water-to-air, refrigerant-to-air, refrigerant-to-water, and air-to-air types. These types are covered in other sections of this guide.

Relevant Handbook Chapter:

ASHRAE. 1985. Chapter 3, "Heat transfer." *ASHRAE handbook–1985 fundamentals*. Atlanta: American Society of Heating, Refrigerating, and Air-Conditioning Engineers, Inc.

Key References:

Semi-Theoretical Algorithms
DOE-2. 1982. *Engineer's manual, Version 2.1A*. Lawrence Berkeley Laboratory, Berkeley, California, and Los Alamos National Laboratory, Los Alamos, New Mexico. NTIS, Reference DE83004575. pp. V.88–V.92.
Silver, S.C.; Jones, J.W.; Peterson, J.L.; Milbitz, A.; and Hunn, B.D. 1988. *CBS/ICE program user's guide*. Center for Energy Studies, The University of Texas at Austin, Conservation and Solar Research Report No. 7, June. pp. 43–44.
Semi-Theoretical Models
Clarke, J.A. 1985. *Energy simulation in building design*. Boston: Adam Hilger Ltd. pp. 252–254.*
Gajanana, B.C. 1979. *Computer simulation and application studies of a space conditioning system with a diesel driven chiller and dual tank thermal storage*. Ph.D. Thesis, Drexel University, Michigan. pp. 40–42.
Hanby, V.I., and Clarke, J.A. 1987. *Catalogue of available HVAC plant component models in the U.K.* Report to SERC GR/D/07459. University of Strathclyde, Glasgow. Code A5.*
Jalali, E.A. 1979. *Cooling tower analyses and economic evaluation*. M.Sc. Thesis, California State University, Long Beach. pp. 47–51.
Stoecker, W.F. 1980. *Design of thermal systems*, second edition. Toronto: McGraw-Hill Book Company. pp. 76–88.
Thal-Larsen, H. 1960. "Dynamics of heat exchangers and their models." *Journal of Basic Engineering, ASME Transactions*, Vol. 82, Series D, No. 2 (June). pp. 489–504.*
TRNSYS. 1981. *A transient system simulation program: Reference manual*. Solar Energy Laboratory, University of Wisconsin–Madison. pp. 4.5-1–4.5-6.
Fundamental Principle Models
Afgan, N.H., and Schlunder, E.U. 1974. *Heat exchangers: Design and theory sourcebook*. Washington, D.C.: Scripta Book Company.*
Enns, M. 1962. "Comparison of dynamic models of a superheater." *Journal of Heat Transfer, ASME Transactions*, No. 4 (November). pp. 375–385.*
Finlay, I.C. 1964. *The dynamic response of heat exchangers to disturbances in flowrate*. National Engineering Laboratory, Report No. 170.*
Hopkinson, A. 1962. *Two models for the dynamics of a crossflow heat exchanger*. United Kingdom Atomic Energy Authority–Reactor Group AEEW-R 258.*
Kays, W.M., and London, A.L. 1964. *Compact heat exchangers*, second edition. Toronto: McGraw-Hill Book Company.*
Kreith, F. 1969. *Principles of heat transfer*, second edition. Scranton, Pennsylvania: International Textbook Company. pp. 483–508.
McQuiston, F.C., and Parker, J.D. 1982. *Heating, ventilating, and air conditioning*. Second Edition. New York: John Wiley and Sons. pp. 469–520.
Rohsenow, W.M., and Hartnett, J.P. 1973. *Handbook of heat transfer*. Toronto: McGraw-Hill Book Company. pp. 18-1–18-112.

Related References:

Rich, D.G. 1973. "The effect of fin spacing on the heat transfer and friction performance of multi-row, smooth plate fin-and-tube heat exchangers." *ASHRAE Transactions*, Vol. 79, Part 2. pp. 137–145.

Rich, D.G. 1975. "The effect of the number of tube rows on heat transfer performance of smooth plate fin-and-tube heat exchangers." *ASHRAE Transactions*, Vol. 81, Part 1. pp. 307–319.

Samuels, D.W. 1982. *The effects of various system parameters on the predictability of heat exchanger performance*. M.Sc. Thesis, University of Kentucky.

7.7 CONTROLS

7.7.1 Valves

Equipment Description:

A valve is a device, operable manually or automatically, that acts as a variable orifice for regulating or varying the flow of steam or water. Types of valves include single-seated devices, double-seated (balanced) devices, three-way mixing devices, three-way diverting devices, and butterfly devices. The performance of a valve is described by its flow characteristics as it operates through its stroke, based on a constant pressure drop. Three common flow characteristics are equal percentage, linear, and quick-opening. Valves that regulate the flow of refrigerant are covered in another section of this guide.

Relevant Handbook Chapter:

ASHRAE. 1987. Chapter 51, "Automatic control." *ASHRAE handbook–1987 HVAC systems and applications*. Atlanta: American Society of Heating, Refrigerating, and Air-Conditioning Engineers, Inc.

Key References:

Semi-Theoretical Algorithms

ASHRAE. 1976. *Energy calculations 2–Procedures for simulating the performance of components and systems for energy calculations*. Atlanta: American Society of Heating, Refrigerating, and Air-Conditioning Engineers, Inc. pp. 62–63.

Semi-Theoretical Models

Clark, D.R. 1985. "Damper or valve." *HVACSIM+ building systems and equipment simulation program: Reference manual*. NBSIR 84-2996, National Bureau of Standards, Washington, D.C. pp. 41–44.

Clark, D.R. 1985. "Linear valve with pneumatic actuator." *HVACSIM+ building systems and equipment simulation program: Reference manual*. NBSIR 84-2996, National Bureau of Standards, Washington, D.C. pp. 51–53.

Clark, D.R. 1985. "Three-way valve with actuator." *HVACSIM+ building systems and equipment simulation program: Reference manual*. NBSIR 84-2996, National Bureau of Standards, Washington, D.C. pp. 69–71.

Clark, D.R.; Hurley, C.W.; and Hill, C.R. 1985. "Dynamic models for HVAC system components." *ASHRAE Transactions*, Vol. 91, Part 1B. pp. 737–750.

Hamilton, D.C.; Leonard, R.G.; and Pearson, J.T. 1974. "Dynamic response characteristics of a discharge air temperature control system at near full and part heating load." *ASHRAE Transactions*, Vol. 80, Part 1. pp. 181–194.

Hamilton, D.C.; Leonard, R.G.; and Pearson, J.T. 1977. "A system model for a discharge air temperature control system." *ASHRAE Transactions*, Vol. 83, Part 1. pp. 251–268.

Hill, C.R. 1985. "Simulation of a multizone air handler." *ASHRAE Transactions*, Vol. 91, Part 1B. pp. 752–765.

Laitinen, A., and Virtanen, M. 1987. *Flow control devices*. International Energy Agency, Annex 10, System Simulation, Report AN10-871211-17-01, University of Liege, Belgium.

Mehta, D.P. 1984. "Modeling of environmental control components." *Proceedings of the Workshop on HVAC Controls Modeling and Simulation*. Atlanta.

Fundamental Principle Algorithms

DOE-2. 1982. *Engineer's manual, Version 2.1A*. Lawrence Berkeley Laboratory, Berkeley, California, and Los Alamos National Laboratory, Los Alamos, New Mexico. NTIS, Reference DE83004575. p. V.213.

Silver, S.C.; Jones, J.W.; Peterson, J.L.; Milbitz, A.; and Hunn, B.D. 1988. *CBS/ICE program user's guide*. Center for Energy Studies, The University of Texas at Austin, Conservation and Solar Research Report No. 7, June. pp. 49–50.

Fundamental Principle Models

TRNSYS. 1981. *A transient system simulation program: Reference manual*. Solar Energy Laboratory, University of Wisconsin–Madison. pp. 4.11-1–4.11-9.

Zhang, X., and Warren, M.L. 1988. "Use of a general control simulation program." *ASHRAE Transactions*, Vol. 94, Part 1.

Related References:

Ast, H. 1988. *Thermostatic valve*. International Energy Agency, Annex 10, System Simulation, Report AN10-880419-02, University of Liege, Belgium.

7.7.2 Dampers

Equipment Description:

A damper is a device, operable manually or automatically, that acts as a variable orifice for regulating or varying the flow of air. Types of dampers include single-blade devices, parallel-blade devices, and opposed-blade devices. The performance of a damper is described by its flow characteristics as it operates through its stroke, based on a constant pressure drop.

Relevant Handbook Chapter:

ASHRAE. 1987. Chapter 51, "Automatic control." *ASHRAE handbook–1987 HVAC systems and applications*. Atlanta: American Society of Heating, Refrigerating, and Air-Conditioning Engineers, Inc.

Key References:

Semi-Theoretical Models

Clark, D.R. 1985. "Damper or valve." *HVACSIM+ building systems and equipment simulation program: Reference manual*. NBSIR 84-2996, National Bureau of Standards, Washington, D.C. pp. 41–44.

Clark, D.R. 1985. "Mixing dampers and merge." *HVACSIM+ building systems and equipment simulation program: Reference manual*. NBSIR 84-2996, National Bureau of Standards, Washington, D.C. pp. 79–80.

Clark, D.R.; Hurley, C.W.; and Hill, C.R. 1985. "Dynamic models for HVAC system components." *ASHRAE Transactions*, Vol. 91, Part 1B. pp. 737–750.

Hanby, V.I. 1988. *Dampers*. International Energy Agency, Annex 10, System Simulation, Report AN10-880202-03, University of Liege, Belgium.

Hill, C.R. 1985. "Simulation of a multizone air handler." *ASHRAE Transactions*, Vol. 91, Part 1B. pp. 752–765.

Mehta, D.P. 1984. "Modeling of environmental control components." *Proceedings of the Workshop on HVAC Controls Modeling and Simulation*. Atlanta.

7.7.3 Operators

Equipment Description:

An operator is the link between the controller and a valve or damper assembly. It can use electricity (solenoids or motors), compressed air, or hydraulic fluid to power the motion of the valve stem or damper linkage through its operating range. In many cases, the operator amplifies the input signal.

Relevant Handbook Chapter:

ASHRAE. 1987. Chapter 51, "Automatic control." *ASHRAE handbook–1987 HVAC systems and applications*. Atlanta: American Society of Heating, Refrigerating, and Air-Conditioning Engineers, Inc.

Key References:

Semi-Theoretical Models

Brandt, S.G., and Shavit, G. 1984. "Simulation of the PID algorithm for direct digital control applications." *Proceedings of the Workshop on HVAC Controls Modeling and Simulation.* Atlanta.*

Clark, D.R. 1985. "Linear valve with pneumatic actuator." *HVACSIM+ building systems and equipment simulation program: Reference manual*. NBSIR 84-2996, National Bureau of Standards, Washington, D.C. pp. 51–53.*

Clark, D.R. 1985. "Three-way valve with actuator." *HVACSIM+ building systems and equipment simulation program: Reference manual*. NBSIR 84-2996, National Bureau of Standards, Washington, D.C. pp. 69–71.*

Laitinen, A., and Virtanen, M. 1987. *Flow control devices*. International Energy Agency, Annex 10, System Simulation, Report AN10-871211-17-01, University of Liege, Belgium.*

Shavit, G., and Brandt, S.G. 1987. "Dynamic performance of a discharge air-temperature system with a P-I controller." *ASHRAE Journal*, No. 11 (November).*

Fundamental Principle Models

Hamilton, D.C.; Leonard, R.G.; and Pearson, J.T. 1977. "A system model for a discharge air temperature control system." *ASHRAE Transactions*, Vol. 83, Part 1. pp. 251–268.*

Mehta, D.P. 1984. "Modeling of environmental control components." *Proceedings of the Workshop on HVAC Controls Modeling and Simulation*. Atlanta.*

Related References:

Zhang, X., and Warren, M.L. 1988. "Use of a general control simulation program." *ASHRAE Transactions*, Vol. 94, Part 1.

7.7.4 Sensors

Equipment Description:

A sensor is a component of a control system that measures the value of a controlled variable, such as temperature. A change in the controlled variable produces a change in some physical or electrical property of the primary sensing element. This change in property is then available for translation or amplification as a mechanical or electrical output signal. Sensor types include temperature-sensing elements, humidity sensors (hygrometers), pressure transducers, and fluid flow-rate sensors.

Relevant Handbook Chapter:

ASHRAE. 1987. Chapter 51, "Automatic control." *ASHRAE handbook–1987 HVAC systems and applications*. Atlanta: American Society of Heating, Refrigerating, and Air-Conditioning Engineers, Inc.

Key References:

Semi-Theoretical Models

Brandt, S.G., and Shavit, G. 1984. "Simulation of the PID algorithm for direct digital control applications." *Proceedings of the Workshop on HVAC Controls Modeling and Simulation.* Atlanta.*

Clark, D.R. 1985. "Temperature sensor." *HVACSIM+ building systems and equipment simulation program: Reference manual.* NBSIR 84-2996, National Bureau of Standards, Washington, D.C. pp. 47–48.*

Clark, D.R.; Hurley, C.W.; and Hill, C.R. 1985. "Dynamic models for HVAC system components." *ASHRAE Transactions,* Vol. 91, Part 1B. pp. 737–750.*

Hamilton, D.C.; Leonard, R.G.; and Pearson, J.T. 1977. "A system model for a discharge air temperature control system." *ASHRAE Transactions,* Vol. 83, Part 1. pp. 251–268.*

Hill, C.R. 1985. "Simulation of a multizone air handler." *ASHRAE Transactions,* Vol. 91, Part 1B. pp. 752–765.*

Kao, J.Y.; Sushinsky, G.; Didion, D.A.; Mastascusa, E.J.; and Chi, J. 1983. *Low-voltage room thermostat performance.* National Bureau of Standards Building Science Series 150.*

Mehta, D.P. 1984. "Modeling of environmental control components." *Proceedings of the Workshop on HVAC Controls Modeling and Simulation.* Atlanta.*

Shavit, G., and Brandt, S.G. 1987. "Dynamic performance of a discharge air-temperature system with a P-I controller." *ASHRAE Journal,* No. 11 (November).*

Fundamental Principle Models

Clarke, J.A. 1985. *Energy simulation in building design.* Boston: Adam Hilger Ltd. pp. 270–272.*

7.7.5 Controllers

Equipment Description:

A controller is a device that receives input signals from one or more sensors; compares these signals with desired control conditions (e.g., setpoints, limits); and generates one or more output signals to cause operators to provide control actions on controlled devices, such as valves or dampers. Controllers include the following types: electric/electronic devices; indicating or recording devices; pneumatic receiver devices; direct digital control (DDC) devices; and thermostats. Thermostats are a special class of controllers because they combine the functions of a sensor and a controller.

Relevant Handbook Chapter:

ASHRAE. 1987. Chapter 51, "Automatic control." *ASHRAE handbook–1987 HVAC systems and applications.* Atlanta: American Society of Heating, Refrigerating, and Air-Conditioning Engineers, Inc.

Key References:

Semi-Theoretical Algorithms

Ford, J.W. 1980. "A direct digital control system in Cornwall, Ontario." *ASHRAE Transactions,* Vol. 86, Part 1. pp. 895–906.*

Semi-Theoretical Models

Clark, D.R. 1985. "Proportional-integral controller." *HVACSIM+ building systems and equipment simulation program: Reference manual.* NBSIR 84-2996, National Bureau of Standards, Washington, D.C. pp. 49–50.*

Clark, D.R. 1985. " 'Sticky' proportional controller." *HVACSIM+ building systems and equipment simulation program: Reference manual.* NBSIR 84-2996, National Bureau of Standards, Washington, D.C. pp. 77–78.

Clark, D.R. 1985. "High or low-limit controller." *HVACSIM+ building systems and equipment simulation program: Reference manual.* NBSIR 84-2996, National Bureau of Standards, Washington, D.C. p. 86.

Clarke, J.A. 1985. *Energy simulation in building design.* Boston: Adam Hilger Ltd. p. 270.*

EMPS-2.1. 1988. *Computer program for residential building energy analysis–Engineering* *manual*. Electric Power Research Institute, Palo Alto, California. pp. 3-2-3-3.*

Farris, D.R., and McDonald, T.E. 1980. "Adaptive optimal control–An algorithm for direct digital control." *ASHRAE Transactions*, Vol. 86, Part 1. pp. 880–893.*

Hamilton, D.C.; Leonard, R.G.; and Pearson, J.T. 1977. "A system model for a discharge air temperature control system." *ASHRAE Transactions*, Vol. 83, Part 1. pp. 251–268.*

Hill, C.R. 1985. "Simulation of a multizone air handler." *ASHRAE Transactions*, Vol. 91, Part 1B. pp. 752–765.*

Laitinen, A., and Virtanen, M. 1987. *Flow control devices*. International Energy Agency, Annex 10, System Simulation, Report AN10-871211-17-01, University of Liege, Belgium.*

Mehta, D.P. 1984. "Modeling of environmental control components." *Proceedings of the Workshop on HVAC Controls Modeling and Simulation*. Atlanta.*

Nguyen, H.V., and Goldschmidt, V. 1983. "Modeling of a residential thermostat and the duty cycle of a compressor-driven HVAC system." *ASHRAE Transactions*, Vol. 89, Part 2A. pp. 361–372.*

Shavit, G., and Brandt, S.G. 1987. "Dynamic performance of a discharge air-temperature system with a P-I controller." *ASHRAE Journal*, No. 11 (November).*

Stoecker, W.F.; Rosario, L.A.; Heidenreich, M.E.; and Phelen, T.R. 1978. "Stability of an air-temperature control loop." *ASHRAE Transactions*, Vol. 84, Part 1. pp. 35–53.*

Zhang, X., and Warren, M.L. 1988. "Use of a general control simulation program." *ASHRAE Transactions*, Vol. 94, Part 1.*

Fundamental Principle Algorithms

DOE-2. 1982. *Engineer's manual, Version 2.1A*. Lawrence Berkeley Laboratory, Berkeley, California, and Los Alamos National Laboratory, Los Alamos, New Mexico. NTIS, Reference DE83004575. pp. V.205–V.206.

Fundamental Principle Models

Brandt, S.G., and Shavit, G. 1984. "Simulation of the PID algorithm for direct digital control applications." *Proceedings of the Workshop on HVAC Controls Modeling and Simulation*. Atlanta.*

TRNSYS. 1981. "On/off differential controller with hysteresis." *A Transient System Simulation Program: Reference Manual*. Solar Energy Laboratory, University of Wisconsin–Madison. pp. 4.2-1–4.2-4.

TRNSYS. 1981. "Three-stage room thermostat." *A Transient System Simulation Program: Reference Manual*. Solar Energy Laboratory, University of Wisconsin–Madison. pp. 4.8-1–4.8-4.

Related References:

Chalifoux, A.T. 1986. *A generalized means of simulating HVAC distribution systems on a computer*. M.Sc. Thesis, University of Illinois at Urbana–Champaign.

Kao, J.Y.; Sushinsky, G.; Didion, D.A.; Mastascusa, E.J.; and Chi, J. 1983. *Low-voltage room thermostat performance*. National Bureau of Standards Building Science Series 150.

McBride, M.F. 1979. "Measurement of residential thermostat dynamics for predicting transient performance." *ASHRAE Transactions*, Vol. 85, Part 1. pp. 684–694.

7.8 FUNDAMENTALS

7.8.1 Psychrometrics

ASHRAE. 1977. *ASHRAE brochure on psychrometry*. Atlanta: American Society of Heating, Refrigerating, and Air-Conditioning Engineers, Inc.

ASHRAE. 1985. Chapter 6, "Psychrometrics." *ASHRAE handbook–1985 fundamentals*. Atlanta: American Society of Heating, Refrigerating, and Air-Conditioning Engineers, Inc.

Carroll, J.T. 1984. *Computer simulation of heat pump systems in residential applications*. M.Sc. Thesis, Department of Mechanical and Aerospace Engineering, North Carolina State University. pp. 19–25.

DOE-2. 1982. *Engineer's manual, Version 2.1A*. Lawrence Berkeley Laboratory, Berkeley,

California, and Los Alamos National Laboratory, Los Alamos, New Mexico. NTIS, Reference DE83004575. pp. IV.211–IV.214.

Ellison, R.D., and Creswick, F.A. 1978. *A computer simulation of steady-state performance of air-to-air heat pumps.* ORNL/CON-16, Oak Ridge National Laboratory. pp. 78–79.

HAP. 1987. *E20-II–Hourly analysis program: User's manual, Version 1.1.* Carrier Corporation, Syracuse, New York.

Howell, R.H.; Sauer, H.J.; and Ganesh, R. 1987. "Comparison of two control strategies to simulate part-load performance of a simple air-conditioning system." *ASHRAE Transactions*, Vol. 93, Part 2.

Marcev, C.L.; Smith, C.R.; and Bruns, D.D. 1984. "Supervisory control and system optimization of chiller-cooling tower combinations via simulation using primary equipment component models: Part I. Primary equipment component models." *Proceedings of the Workshop on HVAC Control Modeling and Simulation.* Atlanta.

McQuiston, F.C., and Parker, J.D. 1982. *Heating, ventilating, and air conditioning.* Second Edition. New York: John Wiley and Sons. pp. 11–45.

Threlkeld, J.L. 1970. *Thermal environmental engineering,* second edition. Englewood Cliffs, New Jersey: Prentice-Hall, Inc. pp. 165–213.

7.8.2 Equation Fitting

ASHRAE. 1976. *Energy calculations 2–Procedures for simulating the performance of components and systems for energy calculations.* Atlanta: American Society of Heating, Refrigerating, and Air-Conditioning Engineers, Inc. p. 7–29.

Leah, R.L. 1983. *Quadratic search algorithms for optimization of thermal system models.* M.Sc. Thesis, University of Illinois at Urbana–Champaign.

Stoecker, W.F. 1980. *Design of thermal systems,* second edition. Toronto: McGraw-Hill Book Company. pp. 50–70.

Wright, J.A. 1986. "The application of the least squares curve fitting method to the modelling of HVAC component performance." *Proceedings of the Second International Conference on System Simulation in Buildings.* Liege, Belgium.

7.8.3 Heat Transfer in Pipes and Ducts

ASHRAE. 1976. *Energy calculations 2–Procedures for simulating the performance of components and systems for energy calculations.* Atlanta: American Society of Heating, Refrigerating, and Air-Conditioning Engineers, Inc. pp. 64.

ASHRAE. 1985. Chapter 3, "Heat transfer." *ASHRAE handbook–1985 fundamentals.* Atlanta: American Society of Heating, Refrigerating, and Air-Conditioning Engineers, Inc.

Clark, D.R. 1985. *HVACSIM+ building systems and equipment simulation program: Reference manual.* NBSIR 84-2996, National Bureau of Standards, Washington, D.C. pp. 33–36.

Clark, D.R.; Hurley, C.W.; and Hill, C.R. 1985. "Dynamic models for HVAC system components." *ASHRAE Transactions*, Vol. 91, Part 1B. pp. 737–750.

EMPS-2.1. 1988. *Computer program for residential building energy analysis–Engineering manual.* Electric Power Research Institute, Palo Alto, California. pp. 3-4-3-7.

Hamilton, D.C.; Leonard, R.G.; and Pearson, J.T. 1977. "A system model for a discharge air temperature control system." *ASHRAE Transactions*, Vol. 83, Part 1. pp. 251–268.

Hanby, V.I., and Clarke, J.A. 1987. *Catalogue of available HVAC plant component models in the U.K.* Report to SERC GR/D/07459. University of Strathclyde, Glasgow. Code L1.

IEA. 1986. *System simulation: Synthesis report for three years of activities (1983-1985)–The simulation exercises, Level 1.* Document AN10 860429-01. IEA Annex 10, University of Liege, Belgium.

Jones, J.W.; Jones, C.D.; Sepsy, C.F.; and Crall, G.C. 1975. "Simulation of a dual duct, reheat air-handling system." *ASHRAE Transactions*, Vol. 81, Part 1. pp. 463–474.

Kreith, F. 1969. *Principles of heat transfer,* second edition. Scranton, Pennsylvania: International Textbook Company. pp. 483–516.

Malmstrom, T.G. 1988. *Piping.* International Energy Agency, Annex 10, System Simulation, Report AN10-880603-02, University of Liege, Belgium.

Malmstrom, T.G., and Olsson, L.G. 1988. *Ducts*. International Energy Agency, Annex 10, System Simulation, Report AN10-880603-06, University of Liege, Belgium.

McQuiston, F.C., and Parker, J.D. 1982. *Heating, ventilating, and air conditioning*. Second Edition. New York: John Wiley and Sons. pp. 122–126, 174–175.

Rohsenow, W.M., and Hartnett, J.P. 1973. *Handbook of heat transfer*. Toronto: McGraw-Hill Book Company. pp. 18-1–18-113.

Silver, S.C.; Jones, J.W.; Peterson, J.L.; Milbitz, A.; and Hunn, B.D. 1988. *CBS/ICE program user's guide*. Center for Energy Studies, The University of Texas at Austin, Conservation and Solar Research Report No. 7, June. pp. 157–159.

Stoecker, W.F. 1958. *Refrigeration and air conditioning*. New York: McGraw-Hill Book Company. pp. 141.

Tobias, J.R. 1973. "Simplified transfer function for temperature response of fluids flowing through coils, pipes or ducts." *ASHRAE Transactions*, Vol. 79, Part 2. pp. 19–22.

TrakLoad. 1986. *Energy audit system: Reference manual, Version 3.1*. Morgan Systems Corporation, Berkeley, California. pp. 3-69–3-70.

TRNSYS. 1981. *A transient system simulation program: Reference manual*. Solar Energy Laboratory, University of Wisconsin–Madison. pp. 4.31-1–4.31-3.

7.8.4 Pressure Drop in Pipes and Ducts

AMCA. 1987. *Air Systems*. Arlington Heights VA: Air Movement and Control Association, Inc.

Arneodo, P., and Mazza, A. 1988. *Air filters*. International Energy Agency, Annex 10, System Simulation, Report AN10-880520-08, University of Liege, Belgium.

ASHRAE. 1985. Chapter 2, "Fluid flow." *ASHRAE handbook–1985 fundamentals*. Atlanta: American Society of Heating, Refrigerating, and Air-Conditioning Engineers, Inc.

Clark, D.R. 1985. *HVACSIM+ building systems and equipment simulation program: Reference manual*. NBSIR 84-2996, National Bureau of Standards, Washington, D.C. pp. 33–40, 45–46, 87–88.

Clark, D.R.; Hurley, C.W.; and Hill, C.R. 1985. "Dynamic models for HVAC system components." *ASHRAE Transactions*, Vol. 91, Part 1B. pp. 737–750.

Fischer, S.K., and Rice, C.K. 1983. *The Oak Ridge heat pump models: I. A steady-state computer design model for air-to-air heat pumps*. ORNL/CON-80/R1, Oak Ridge National Laboratory. pp. 65–72.

Hamilton, D.C.; Leonard, R.G.; and Pearson, J.T. 1974. "Dynamic response characteristics of a discharge air temperature control system at near full and part heating load." *ASHRAE Transactions*, Vol. 80, Part 1. pp. 181–194.

Hamilton, D.C.; Leonard, R.G.; and Pearson, J.T. 1977. "A system model for a discharge air temperature control system." *ASHRAE Transactions*, Vol. 83, Part 1. pp. 251–268.

Hanby, V.I., and Clarke, J.A. 1987. *Catalogue of available HVAC plant component models in the U.K*. Report to SERC GR/D/07459. University of Strathclyde, Glasgow. Codes L1, L2, L3, L4.

Hill, C.R. 1985. "Simulation of a multizone air handler." *ASHRAE Transactions*, Vol. 91, Part 1B. pp. 752–765.

Malmstrom, T.G. 1988. *Piping*. International Energy Agency, Annex 10, System Simulation, Report AN10-880603-02, University of Liege, Belgium.

Malmstrom, T.G., and Olsson, L.G. 1988. *Ducts*. International Energy Agency, Annex 10, System Simulation, Report AN10-880603-06, University of Liege, Belgium.

Mehta, D.P. 1984. "Modeling of environmental control components." *Proceedings of the Workshop on HVAC Controls Modeling and Simulation*. Atlanta.

McQuiston, F.C., and Parker, J.D. 1982. *Heating, ventilating, and air conditioning*. Second Edition. New York: John Wiley and Sons. pp. 290–295, 386–401.

Silver, S.C.; Jones, J.W.; Peterson, J.L.; Milbitz, A.; and Hunn, B.D. 1988. *CBS/ICE program user's guide*. Center for Energy Studies, The University of Texas at Austin, Conservation and Solar Research Report No. 7, June. pp. 157–159.

Stoecker, W.F. 1980. *Design of thermal systems*, second edition. Toronto: McGraw-Hill Book Company. p. 97.

8. REFERENCES

ASHRAE. 1976. *Energy calculations 2–Procedures for simulating the performance of components and systems for energy calculations*. Atlanta: American Society of Heating, Refrigerating, and Air-Conditioning Engineers, Inc.

ASHRAE. 1985. *ASHRAE handbook–1985 fundamentals*. Atlanta: American Society of Heating, Refrigerating, and Air-Conditioning Engineers, Inc.

ASHRAE. 1986. *ASHRAE handbook–1986 refrigeration*. Atlanta: American Society of Heating, Refrigerating, and Air-Conditioning Engineers, Inc.

ASHRAE. 1987. *ASHRAE handbook–1987 systems and applications*. Atlanta: American Society of Heating, Refrigerating, and Air-Conditioning Engineers, Inc.

ASHRAE. 1988. *ASHRAE handbook–1988 equipment*. Atlanta: American Society of Heating, Refrigerating, and Air-Conditioning Engineers, Inc.

ASHRAE ANNOTATED BIBLIOGRAPHY UPDATE
(Literature Classification Form)

Reference Identification:

Authors
(Surname, Initials):
_____
_____
_____
_____

Article Title:
(Sentence-style
capitalization)
_____
_____
_____

Journal or Book Title:
_____
_____
_____
_____

If Journal (Vol. No., Pages) _____

If Book (City: Publisher) _____

Date of Publication: _____

Reference Description:
(terminology described in Chapter 1 of this guide)

Equipment Type:
_____
_____
_____

Equipment Description:
(Use additional sheets
if necessary)
_____
_____
_____
_____
_____
_____
_____

Relevant ASHRAE Handbook:_____
(Volume and Chapter)
_____
_____

(OVER)

Reference Classification:
(terminology described in Chapter 1 of this guide)

Is this a *Key* **Reference**? [ ] *or* a *Related* **Reference**? [ ] (Check only one.)

If this is a *Key* Reference, what types of algorithms or models does it contain? (Check all categories that apply to this reference.)

	Algorithms	Models	Transient? Yes	Transient? No
Empirical	[ ]	[ ]	[ ]	[ ]
Semi-Empirical	[ ]	[ ]	[ ]	[ ]
Semi-Theoretical	[ ]	[ ]	[ ]	[ ]
Fundamental Principle	[ ]	[ ]	[ ]	[ ]

Submitter Identification:

Name: _____

Address: _____
_____
_____

Phone: _____
Fax: _____

Please submit this form, along with a complete copy of the reference, to:

William W. Seaton
Manager of Research
ASHRAE
1791 Tullie Circle, NE
Atlanta GA 30329